A Short History of Biology

ISAAC ASIMOV is Associate Professor of Biochemistry at Boston University School of Medicine. He obtained his Ph.D. in chemistry from Columbia University in 1948 and has written voluminously on a number of branches of science and mathematics.

In recent years, he has grown particularly interested in the history of science, as is evidenced not only by this book but by his larger *Asimov's Biographical Encyclopedia of Science and Technology.*

Professor Asimov is married and lives in West Newton, Massachusetts, with his wife and two children.

Books on Biology by Isaac Asimov

I am grateful to Professor Everett Mendelsohn of Harvard University and Sir Gavin de Beer, director of the Natural History Department of the British Museum, for their numerous helpful comments in connection with the manuscript of this book. They are, of course, not to be held responsible for any errors of omission or commission that remain.

A SHORT HISTORY
OF BIOLOGY

by Isaac Asimov

GREENWOOD PRESS, PUBLISHERS
WESTPORT. CONNECTICUT

Library of Congress Cataloging in Publication Data

Asimov, Isaac, 1920-
  A short history of biology.

    Reprint of the ed. published by Natural History Press,
Garden City, N.Y.
    Includes index.
    1. Biology--History. I. Title. [DNLM: 1. Biology
--History. QH305 A832s 1964a]
[QH305.A8 1980]   574'.09   80-15464
ISBN 0-313-22583-4 (lib. bdg.)

To

The American Museum of Natural History

WHERE BIOLOGY HAS COME ALIVE TO ME

ON A NUMBER OF OCCASIONS

Originally published for the American Museum of Natural History.
American Museum Science Books Edition: 1964.

A *Short History of Biology* was published simultaneously in a hard-
bound edition by The Natural History Press.

The line illustrations for this book were prepared by the Graphic
Arts Division of The American Museum of Natural History.

Reprinted by arrangement with Doubleday & Company, Inc.

Reprinted in 1980 by Greenwood Press, a division of
Congressional Information Service, Inc.
88 Post Road West, Westport, Conn. 06881

Printed in the United States of America

P

Contents

List of Illustrations

CHAPTER 1

Ancient Biology

The Beginning of Science

Biology is the study of living organisms and as soon as man's mind developed to the point where he was conscious of himself as an object different from the unmoving and unfeeling ground upon which he stood, a form of biology began. For uncounted centuries, however, biology was not in the form we would recognize as a science. Men were bound to attempt to cure themselves and others of ailments, to try to alleviate pain, restore health, and ward off death. They did so, at first, by magical or by religious rites; attempting to force or cajole some god or demon into altering the course of events.

Again, men could not help but observe the living machinery of the animal organism, whenever a creature was cut up by butchers for food or by priests for sacrifice. And yet such attention as was devoted to the detailed features of organs was not with the intent of studying their workings but for the purpose of learning what information they might convey concerning the future. The early anatomists were the diviners who forecast the fate of kings and nations by the shape and appearance of the liver of a ram.

Undoubtedly, much useful information was gathered over the ages, even under the overpowering influence of superstition. The men who embalmed mummies so skillfully in ancient Egypt had to have a working knowledge of human anatomy. The code of Hammurabi, dating back in Babylonian history to perhaps 1920 B.C., included detailed regulations of the medical profession and there were physicians of that day whose knowledge, gleaned from generations of practical observation, must have been both useful and helpful.

Nevertheless, as long as men believed the universe to be under the absolute dominion of capricious demons; as long as they felt the natural to be subordinate to the supernatural, progress in science had to be glacially slow. The best minds would naturally devote themselves not to a study of the visible world, but to attempts to reach, through inspiration or revelation, an understanding of the invisible and controlling world beyond.

To be sure, individual men must now and then have rejected this view and concentrated on the study of the world as it was revealed through the senses. These men, however, lost and submerged in a hostile culture, left their names unrecorded and their influence unfelt.

It was the ancient Greeks who changed that. They were a restless people, curious, voluble, intelligent, argumentative, and, at times, irreverent. The vast majority of Greeks, like all other peoples of the time and of earlier centuries, lived in the midst of an invisible world of gods and demigods. If their gods were far more attractive than the heathen deities of other nations, they were no less childish in their motivations and responses. Disease was caused by the arrows of Apollo, who could be stirred to indiscriminate wrath by some tiny cause and who could be propitiated by sacrifices and appropriate flattery.

But there were Greeks who did not share these views. About 600 B.C., there arose in Ionia (the Aegean coast of what is now Turkey) a series of philosophers, who began a movement that was to change all that. By tradition, the first of these was Thales (640?–546 B.C.).

The Ionian philosophers ignored the supernatural and supposed, instead, that the affairs of the universe followed a fixed and unalterable pattern. They assumed the existence of causality; that is, that every event had a cause, and that a particular cause inevitably produced a particular effect, with no danger of change by a capricious will. A further assumption was that the "natural law" that governed the universe was of such a kind that the mind of man could encompass it and could deduce it from first principles or from observation.

This point of view dignified the study of the universe. It maintained that man could understand the universe and gave the assurance that the understanding, once gained, would be permanent. If one could work out a knowledge of the laws governing the motion of the sun, for instance, one would not need to fear that the knowledge would suddenly become useless when some Phaëthon decided to seize the reins of the sun chariot and lead it across the sky along an arbitrary course.

Little is known of these early Ionian philosophers; their works are lost. But their names survive and the central core of their teachings as well. Moreover the philosophy of "rationalism" (the belief that the workings of the universe could be understood through reason rather than revelation), which began with them, has never died. It had a stormy youth and flickered nearly to extinction after the fall of the Roman Empire, but it never quite died.

Ionia

Rationalism entered biology when the internal machinery of the animal body came to be studied for its own sake, rather than as transmitting devices for divine messages. By tradition, the first man to dissect animals merely to describe what he saw was Alcmaeon (flourished, sixth century B.C.). About 500 B.C., Alcmaeon described the nerves of the eye and studied the structure of the growing chick within the egg. He might thus be considered the first student of *anatomy* (the study of the structure of living organisms) and of *embryology* (the study of organisms before actual birth). Alcmaeon even described the narrow tube that connects the middle ear with the throat. This was lost sight of by later anatomists and was only rediscovered two thousand years later.

The most important name to be associated with the rationalistic beginnings of biology, however, is that of Hippocrates (460?–377? B.C.). Virtually nothing is known about the man himself except that he was born and lived on the island of Cos just off the Ionian coast. On Cos was a temple to Asclepius, the Greek god of medicine. The temple was the nearest equivalent to today's medical school, and to be accepted as a priest there was the equivalent of obtaining a modern medical degree.

Hippocrates' great service to biology was that of reducing Asclepius to a purely honorary position. No god influenced medicine in the Hippocratic view. To Hippocrates, the healthy body was one in which the component parts worked well and harmoniously, whereas a diseased body was one in which they did not. It was the task of the physician to observe closely in order to see where the flaws in the working were, and then to take the proper action to correct those flaws. The proper action did not consist of prayer or sacrifice, of driving out demons or of propitiating gods. It consisted chiefly of allowing the patient to rest, seeing that he was kept clean, had fresh air, and simple wholesome food. Any form of excess was bound to overbalance the body's workings in one respect or another, so there was to be moderation in all things.

In short, the physician's role, in the Hippocratic view, was to let natural law itself effect the cure. The body had self-corrective devices which should be given every opportunity to work. In view of the limited knowledge of medicine, this was an excellent point of view.

Hippocrates founded a medical tradition that persisted for centuries after his time. The physicians of this tradition placed his honored name on their writings so that it is impossible to tell which of the books are actually those of Hippocrates himself. The "Hippocratic oath," for instance, which is still recited by medical graduates at the moment of receiving their degrees, was most certainly not written by him and was, in fact, probably not composed until some six centuries after his time. On the other hand, one of the oldest of the Hippocratic writings deals with the disease epilepsy, and this may very well have been written by Hippocrates himself. If so, it is an excellent example of the arrival of rationalism in biology.

Epilepsy is a disorder of the brain function (still not entirely understood) in which the brain's normal control over the body is disrupted. In milder forms, the victim may misinterpret sense impressions and therefore suffer hallucinations. In the more spectacular forms, the muscles go out of control suddenly; the epileptic falls to the ground and cries out, jerking spasmodically and sometimes doing severe damage to himself.

The epileptic fit does not last long but it is a fearful sight to behold. Onlookers who do not understand the intricacies of the nervous system find it all too easy to believe that if a person moves not of his own volition and in such a way as to harm himself, it must be because some supernatural power has seized control of his body. The epileptic is "possessed"; and the disease is the "sacred disease" because supernatural beings are involved.

In the book *On the Sacred Disease*, written about 400 B.C., possibly by Hippocrates himself, this view is strongly countered. Hippocrates maintained that it was useless, generally, to attribute divine causes to diseases, and that there was no reason to consider epilepsy an exception. Epilepsy, like all other diseases, had a natural cause and a rational treatment. If the cause was not known and the treatment uncertain, that did not change the principle.

All of modern science cannot improve on this view and if one were to insist on seeking for one date, one man, and one book as the beginning of the science of biology, one could do worse than point to the date 400 B.C., the man Hippocrates, and the book *On the Sacred Disease*.

Athens

Greek biology and, indeed, ancient science in general, reached a kind of climax in Aristotle (384–22 B.C.). He was a native of northern Greece and a teacher of Alexander the Great in the latter's youth. Aristotle's great days, however, came in his middle years, when he founded and taught at the famous Lyceum in Athens. Aristotle was the most versatile and thorough of the Greek philosophers. He wrote on almost all subjects, from physics to literature, from politics to biology. In later times, his writings on physics, dealing mainly with the structure and workings of the inanimate universe, were most famous; yet these, as events proved, were almost entirely wrong.

On the other hand, it was biology and, particularly, the study of sea creatures, that was his first and dearest intellectual love. Moreover, it was Aristotle's biological books that proved the best of his scientific writings and yet they were, in later times, the least regarded.

Aristotle carefully and accurately noted the appearance and habits of creatures (this being the study of *natural history*). In the process, he listed about five hundred kinds or "species" of animals, and differentiated among them. The list in itself would be trivial, but Aristotle went further. He recognized that different animals could be grouped into categories and that the grouping was not necessarily done simply and easily. For instance, it is easy to divide land animals into four-footed creatures (beasts); flying, feathered creatures (birds); and a remaining miscellany ("vermin," from the Latin word *vermis* for "worms"). Sea creatures might be all lumped under the heading of "fish." Having done so, however, it is not always easy to tell under which category a particular creature might fit.

Aristotle's careful observations of the dolphin, for instance, made it quite plain that although it was a fishlike creature in superficial appearance and in habitat, it was quite unfishlike in many important respects. The dolphin had lungs and breathed air; unlike fish, it would drown if kept submerged. The dolphin was warm-blooded, not cold-blooded as ordinary fish were. Most important, it gave birth to living young which were nourished before birth by a placenta. In all these respects, the dolphin was similar to hairy warm-blooded animals of the land. These similarities, it seemed to Aristotle, were sufficient to make it necessary to group the cetaceans (the whales, dolphins, and porpoises) with the beasts of the field rather than with the fish of the sea. In this, Aristotle was two thousand years ahead of his time, for cetaceans continued to be grouped with fish throughout ancient and medieval times. Aristotle was quite modern, again, in his division of the scaly fish into two groups, those with bony skeletons and those (like the sharks) with cartilaginous skeletons. This again fits the modern view.

In grouping his animals, and in comparing them with the rest of the universe, Aristotle's neat mind could not resist arranging matters in order of increasing complexity. He saw nature progressing through gradual stages to man, who stands (as it is natural for man to think) at the peak of creation. Thus, one might divide the universe into four kingdoms; the inanimate world of the soil, sea and air; the world of the plants above that; the world of the animals higher still; and the world of man at the peak. The inanimate world exists; the plant world not only exists, it reproduces, too; the animal world not only exists and reproduces, it moves, too; and man not only exists, reproduces and moves, but he can reason, too.

Furthermore, within each world there are further subdivisions. Plants can be divided into the simpler and the more complex. Animals can be divided into those without red blood and those with. The animals without red blood include, in ascending order of complexity, sponges, molluscs, insects, crustaceans, and octopi (according to Aristotle). The animals with red blood are higher on the scale and include fish, reptiles, birds, and beasts.

Aristotle recognized that in this "ladder of life" there were no sharp boundaries and that it was impossible to tell exactly into which group each individual species might fall. Thus very simple plants might scarcely seem to possess any attribute of life. Very simple animals (sponges, for instance) were plantlike, and so on.

Aristotle nowhere showed any traces of belief that one form of life might slowly be converted into another; that a creature high on the ladder might be descended from one lower on the ladder. It is this concept which is the key to modern theories of evolution and Aristotle was not an evolutionist. However, the preparation of a ladder of life inevitably set up a train of thought that was bound, eventually, to lead to the evolutionary concept.

Aristotle is the founder of *zoology* (the study of animals), but as nearly as we can tell from his surviving writings, he rather neglected plants. However, after Aristotle's death, the leadership of his school passed on to his student, Theophrastus (c.380–287 B.C.), who filled in this deficiency of his master. Theophrastus founded *botany* (the study of plants) and in his writings carefully described some five hundred species of plants.

Alexandria

After the time of Alexander the Great and his conquest of the Persian Empire, Greek culture spread rapidly across the Mediterranean world. Egypt fell under the rule of the Ptolemies (descendants of one of the generals of Alexander) and Greeks flocked into the newly founded capital city of Alexandria. There the first Ptolemies founded and maintained the Museum, which was the nearest ancient equivalent of a modern university. Alexandrian scholars are famous for their researches into mathematics, astronomy, geography, and physics. Less important is Alexandrian biology, yet at least two names of the first rank are to be found there. These are Herophilus (flourished about 300 B.C.) and his pupil, Erasistratus (flourished about 250 B.C.).

In Christian times, they were accused of having dissected the human body publicly as a method of teaching anatomy. It is probable they did not do so; more's the pity. Herophilus was the first to pay adequate attention to the brain, which he considered the seat of intelligence. (Alcmaeon and Hippocrates had also believed this, but Aristotle had not. He had felt the brain to be no more than an organ designed to cool the blood.) Herophilus was able to distinguish between sensory nerves (those which receive sensation) and motor nerves (those which induce muscular movement). He also distinguished between arteries and veins, noting that the former pulsated and the latter did not. He described the liver and spleen, the retina of the eye, and the first section of the small intestine (which we now call the "duodenum"). He also described ovaries and related organs in the female and the prostate gland in the male. Erasistratus added to the study of the brain, pointing out the division of the organ into the larger "cerebrum" and the smaller "cerebellum." He particularly noted the wrinkled appearance ("convolutions") of the brain and saw that these were more pronounced in man than in other animals. He therefore connected the convolutions with intelligence.

After such a promising beginning, it seems a pity that the Alexandrian school of biology bogged down, but bog down it did. In fact, all Greek science began to peter out after about 200 B.C. It had flourished for four centuries, but by continuous warfare among themselves, the Greeks had recklessly expended their energies and prosperity. They fell under first Macedonian and then Roman dominion. More and more, their scholarly interests turned toward the study of rhetoric, of ethics, of moral philosophy.

They turned away from natural philosophy—from the rational study of nature that had begun with the Ionians.

Biology, in particular, suffered, for life was naturally considered more sacred than the inanimate universe and therefore less a proper subject for rationalistic study. Dissection of the human body seemed absolutely wrong to many and it either did not take place at all or, if it did, it was soon stopped, first by public opinion, and then by law. In some cases, the objections to dissection lay in the religious belief (by the Egyptians, for instance) that the integrity of the physical body was required for the proper enjoyment of an afterlife. To others, such as the Jews and, later, the Christians, dissection was sacrilegious because the human body was created in the likeness of God, and was therefore holy.

Rome

It came about, therefore, that the centuries during which Rome dominated the Mediterranean world represented one long suspension of biological advance. Scholars seemed content to collect and preserve the discoveries of the past, and to popularize them for Roman audiences. Thus, Aulus Cornelius Celsus (flourished, A.D. 30) collected Greek knowledge into a kind of science-survey course. His sections on medicine survived and were read by Europeans of the early modern era. He thus became more famous as a physician than he truly deserved to be.

The broadening of the physical horizon resulting from Roman conquests made it possible for scholars to collect plants and animals from regions unknown to the earlier Greeks. A Greek physician, Dioscorides (flourished, A.D. 60), who served with the Roman armies, outdid Theophrastus, and described six hundred species of plants. He paid special attention to their medicinal qualities and might thus be considered a founder of *pharmacology* (the study of drugs and medicines).

Even in natural history, however, encyclopedism took over. The Roman name best known in natural history is that of Gaius Plinius Secundus (A.D. 23–79), usually known as Pliny. He wrote a thirty-seven-volume encyclopedia in which he summarized all he could find on natural history among the ancient authors. It was almost all secondary, taken out of the books of others, and Pliny did not always distinguish between the plausible and implausible, so that though his material contains considerable fact (mostly from Aristotle), it also contains a liberal helping of superstition and tall tales (from everywhere else).

Moreover Pliny represents the retreat of the age from rationalism. In dealing with the various species of plants and animals he is very largely concerned with the function of each in connection with man. Nothing exists for its own sake, in his view, but only as food for man, or as a source for medicines, or as a danger designed to strengthen man's muscles and character, or (if all else fails) as a moral lesson. This was a viewpoint to which the early Christians were sympathetic and that, added to the intrinsic interest of his fantasies, accounts in part for the fact that Pliny's volumes survived to modern times.

The last real biologist of the ancient world was Galen (A.D. c.130–c.200), a Greek physician, born in Asia Minor, who practiced in Rome. He had spent his earlier years as a surgeon at the gladiatorial arena and this undoubtedly gave him the opportunity to observe some rough-and-ready human anatomy. However, although the age saw nothing objectionable in cruel and bloody gladiatorial games for the perverted amusement of the populace, it continued to frown at the dissection of dead bodies for scientific purposes. Galen's studies of anatomy had to be based largely on dissections of dogs, sheep, and other animals. When he had the chance, he dissected monkeys for he recognized the manner in which they resembled man.

Galen wrote voluminously and worked out detailed theories on the function of the various organs of the human body. The fact that he was deprived of the chance to study the human body itself and that he lacked modern instruments was the reason most of his theories are not similar to those accepted as true today. He was not a Christian, but he believed strongly in the existence of a single God. Then, too, like Pliny, he believed that everything was made for a purpose, so that he found signs of God's handiwork everywhere in the body. This fitted in with the rising Christian view and helps account for Galen's popularity in later centuries.

CHAPTER 2

Medieval Biology

The Dark Ages

In the latter days of the Roman Empire, Christianity grew to be the dominant religion. When the Empire (or its western regions) was buried under the influx of the Germanic tribes, these, too, were converted to Christianity.

Christianity did not kill Greek science, for that had flickered to near-extinction while Christianity was still but an obscure sect, and, in fact, had showed signs of serious sickness well before the birth of Christ. Nevertheless, the dominance of Christianity worked against the revival of science for many centuries. The Christian viewpoint was quite opposed to that of the Ionian philosophers. To the Christian mind, the important world was not that of the senses, but the "City of God" which could be reached only by revelation and to which the Bible and the writings of the Church fathers and the inspiration of the Church itself were the only sure guides.

The belief in the existence of a natural law that was unchanging and unchangeable gave way to the belief in a world constantly subject to the miraculous interposition of God on behalf of His saints. In fact, it was even felt by some that the study of the things of the world was a devilish device designed to distract the Christian from the proper attention to things of the spirit. Science, from that standpoint, became a thing of evil.

Naturally, this was not the universal view and the light of science maintained a feeble glow amid the shadow of the so-called "Dark Ages." An occasional scholar struggled to keep worldly knowledge alive. For instance, the Englishman, Bede (673–735), preserved what he could of the ancients. Since, however, this consisted largely of scraps of Pliny, what he preserved was not very advanced.

Perhaps, in fact, the light might have faded out after all, had it not been for the Arabs. The Arabs adopted Islam, a religion even newer than Christianity, and preached by Mohammed in the seventh century. They burst out of their arid peninsula at once and flooded over southwestern Asia and northern Africa. By 730, a century after Mohammed, the men of Islam (Moslems) stood at the edge of Constantinople on the east and at the edge of France on the west.

Militarily and culturally, they seemed a dreadful scourge and danger to Christian Europe, but intellectually, they proved, in the long run, to be a boon. Like the Romans, the Arabs were not themselves great scientific originators. Nevertheless, they discovered the work of men such as Aristotle and Galen, translated them into Arabic, preserved them, studied them, and wrote commentaries on them. The most important of the Moslem biologists was the Persian physician, abu-'Ali al-Ḥusayn ibn-Sīna, commonly known by the Latinized version of the last part of his name, Avicenna (980–1037). Avicenna wrote numerous books based on the medical theories of Hippocrates and on the collected material in Celsus' books.

About that time, however, the tide had turned, at least in western Europe. Christian armies had reconquered Sicily which, for a couple of centuries, had been controlled by the Moslems, and were reconquering Spain. Toward the end of the eleventh century, west European armies began to invade the Near East in what are called the Crusades.

Contacts with the Moslems helped make Europeans aware of the fact that the enemy culture was not merely a thing of the devil but that, in some respects, it was more advanced and sophisticated than their own way of life at home. European scholars began to seek after Moslem learning, and projects to translate Arabic books on science flourished. Working in newly reconquered Spain, where the help of Moslem scholars could be counted on, the Italian scholar, Gerard of Cremona (1114–87), translated the works of Hippocrates and Galen, as well as some of the works of Aristotle, into Latin.

A German scholar, Albertus Magnus (1206–80), was one of those who fell in love with the rediscovered Aristotle. His teachings and writings were almost entirely Aristotelian and he helped lay once more a foundation of Greek science on which, at last, more could be built.

One of Albertus' pupils was the Italian scholar, Thomas Aquinas (c.1225–74). He labored to harmonize Aristotelian philosophy and the Christian faith and, by and large, succeeded. Aquinas was a rationalist in that he felt that the reasoning mind was God-created, as was the rest of the universe, and that by true reasoning man could not arrive at a conclusion that was at odds with Christian teaching. Reason was therefore not evil or harmful.

The stage was thus set for a renewal of rationalism.

The Renaissance

In Italy, the practice of dissection was revived in the later Middle Ages. The practice was still in disrepute but there was an important law school at Bologna and it frequently happened that legal questions concerning cause of death might best be decided by a post-mortem study. Once that grew to seem justified it was an easy step to the use of dissection in medical teaching. (Both Bologna and Salerno were noted for their medical schools at the time.)

The revival of dissection did not at once break new ground in biology. At first the primary purpose was to illustrate the works of Galen and Avicenna. The teacher himself was a learned scholar who had studied the books but who felt that the actual dissection was a demeaning job to be left in the hands of an underling. The teacher lectured but did not look to see whether the statements he delivered agreed with the facts, while the underling (no scholar himself) was anxious only to keep from offending the lecturer. The grossest errors were therefore perpetuated, and features that Galen had found in animals and supposed, therefore, to be present in humans were "found" in humans, too, over and over again, though they did not, in fact, exist in humans.

One exception to this sad · situation was the Italian anatomist, Mondino de' Luzzi (1275–1326). At the Bologna medical school, he did his own dissections and, in 1316, wrote the first book to be devoted entirely to anatomy. He is therefore known as the "Restorer of Anatomy." But it was a false dawn. Mondino did not have the courage to break completely with the errors of the past and some of his descriptions must have been based on the evidence of the old books rather than that of his own eyes. Moreover, after his time, the practice of dissection by means of an underling was re-established.

Outside the formal domain of science, however, a new

motivation toward the study of biology was arising in Italy. The period of the rebirth of learning (partly through the rediscovery of the ancient writings and partly through a natural ferment within European culture itself) is referred to as the "Renaissance."

During the Renaissance, a new naturalism in art grew apace. Artists learned how to apply the laws of perspective to make paintings take on a three-dimensional appearance. Once that was done, every effort was made to improve art's mimicry of nature. To make the human figure seem real, one had to study (if one were completely conscientious) not only the contours of the skin itself but also the contours of the muscles beneath the skin; the sinews and tendons; and even the arrangement of the bones. Artists, therefore, could not help but become amateur anatomists.

Perhaps the most famous of the artist-anatomists is the Italian, Leonardo da Vinci (1452–1519), who dissected both men and animals. He had the advantage over ordinary anatomists of being able to illustrate his own findings with drawings of the first quality. He studied (and illustrated) the manner in which the bones and joints were arranged. In doing so, he was the first to indicate accurately how similar the bone arrangements were in the leg of the human and the horse, despite surface differences. This was an example of "homology," which was to unite into firmly knit groups many animals of outwardly diverse appearance and was to help lay further groundwork for theories of evolution.

Leonardo studied and illustrated the mode of working of the eye and the heart; and he pictured plant life as well. Because he was interested in the possibility of devising a machine that would make human flight possible, he studied birds with great attention, drawing pictures of them in flight. All of this, however, he kept in coded notebooks. His contemporaries were unaware of his work, which was discovered only in modern times. He did not, therefore, influence the progress of science, and for his selfish hoarding of knowledge, Leonardo is to be blamed.

As anatomy slowly revived, so did natural history. The fifteenth century had seen an "Age of Exploration" dawn upon Europe, and European ships ranged the coasts of Africa, reached India and the islands beyond, and discovered the Americas. As once before, after the conquests of the Macedonians and the Romans, new and unheard of species of plants and animals roused the curiosity of scholars.

An Italian botanist, Prospero Alpini (1553–1617), served as physician to the Venetian consul in Cairo, Egypt. There he had the opportunity to study the date palm and note that it existed as male and female. Theophrastus had noticed this almost two thousand years before but the fact had been forgotten and the asexuality of plants had been accepted. Alpini was the first European, furthermore, to describe the coffee plant. The natural history of the Renaissance reached its most voluble development with the Swiss naturalist, Konrad von Gesner (1516–65). He was much like Pliny in his wide-ranging interests, his universal curiosity, his tendency to gullibility, and his belief that the mere mass accumulation of excerpts from old books was the way to universal knowledge. In fact, he is sometimes called the "German Pliny."

The Transition

By the early decades of the 1500s, Europe had surged back from the darkness and had reached the limits of Greek biology (and of Greek science in general, in fact). The movement could not progress further, however, unless the scholars of Europe could be made to realize that the Greek books were but a beginning. They had to be discarded, once mastered, and not kept and revered until they became prison walls of the mind. The work of Mondino illustrates how difficult it was to break away from the ancients and move beyond.

Perhaps it took a half-mad boaster to make the break and serve as a living transition to modern times. The one who did so was a Swiss physician named Theophrastus Bombastus von Hohenheim (1493–1541). His father taught him medicine and he himself had a roving foot and a receptive mind. He picked up a great many remedies on his travels that were not known to his stay-at-home contemporaries, and made himself out to be a marvelously learned physician.

He was interested in alchemy, which Europeans had picked up from the Arabs who had, in turn, picked it up from the Alexandrian Greeks. The ordinary alchemist was (when not an outright faker) the equivalent of the modern chemist, but the two most startling goals of alchemy were will-o'-the-wisps never destined to be achieved—at least not by alchemical methods.

Alchemists attempted, first, to find methods of transmuting base metals, such as lead, to gold. Secondly, they sought what was commonly known as the "philosopher's stone"—a dry material supposed by some to be the medium for transmuting metals to gold, and by others to be a universal cure, an elixir of life that was the clue even to immortality.

Hohenheim saw no point in trying to make gold. He believed that the true function of alchemy was to aid the physician in the cure of disease. For this reason, he concentrated on the philosopher's stone which he claimed he had discovered. (He did not hesitate to assert that he would live forever as a result, but, alas, he died before he was fifty of an accidental fall.) Hohenheim's alchemical leanings led him to look to mineral sources for his cures—minerals being the stock in trade of alchemy—and to scorn the botanical medicines that were so in favor with the ancients. He inveighed furiously against the ancients. Celsus' works had just been translated and were the bible of European physicians, but Hohenheim called himself

"Paracelsus" ("better than Celsus") and it is by that vainglorious name that he is known to posterity.

Paracelsus was town physician in Basel in 1527, and to show his opinions as publicly as possible, he burnt copies of the books of Galen and Avicenna in the town square. As a result, his conservative enemies among the medical profession maneuvered him out of Basel, but that did not change his opinions. Paracelsus did not destroy Greek science, or even Greek biology, but his attacks had drawn the attention of scholars. His own theories were not much better than the Greek theories against which he railed so furiously, but it was a time when iconoclasm was necessary and valuable in itself. His loud irreverence against the ancients could not help but shake the pillars of orthodox thought and although Greek science kept its stranglehold on the European mind for a while longer, the hold was weakening perceptibly.

CHAPTER 3

The Birth of Modern Biology

The New Anatomy

The year which is usually considered as marking the beginning of what is called the "Scientific Revolution" is 1543. In that year, Nicolaus Copernicus, a Polish astronomer, published a book describing a new view of the solar system, one in which the sun was at the center, and the earth was a planet that moved in an orbit like any other. This marked the beginning of the end of the old Greek view of the universe (in which the earth was at the center), though a century's hard fighting remained before the victory of the new view was manifest.

In that same year, 1543, a second book was published; one as revolutionary for the biological sciences as Copernicus' book was to prove for the physical sciences. This second book was *De Corporis Humani Fabrica* ("On the Structure of the Human Body") and its author was a Belgian anatomist named Andreas Vesalius (1514–64).

Vesalius was educated in the Netherlands in the strict tradition of Galen, for whom he always retained the greatest respect. However, he traveled to Italy once his education was complete and there he entered a more liberal intellectual atmosphere. He reintroduced Mondino de' Luzzi's old habit of doing his own dissections, and did not allow himself to be influenced by old Greek views when his eyes disagreed with those views.

The book he published, as the result of his observations, was the first accurate book on human anatomy ever presented to the world. It had great advantages over earlier books in two respects. First, it came in an age when printing had been discovered and was in use, so that thousands of copies could be broadcast over Europe. Second, it had illustrations. These illustrations were outstandingly beautiful, many having been done by Jan Stevenzoon van Calcar, a pupil of the artist, Titian. The human body was shown in natural positions and the illustrations of the muscles were particularly good.

Vesalius' life after the appearance of his book was an unhappy one. His views seemed heretical to some and certainly his public dissections, openly advertised by his book, were illegal. He was forced to undertake a pilgrimage to the Holy Land, and was lost in a shipwreck on the way back.

Vesalius' revolution in biology, however, was more immediately effective than Copernicus' revolution in astronomy. What Vesalius' book maintained was not something as incredible (on the surface) as the movement of

the huge earth through space. Rather, it presented in an
attractive manner, the shape and arrangement of organs
that (however much it might run counter to ancient
authority) anyone might see for himself if he troubled
to look.

Greek anatomy was obsolete and a new Italian anatomy flourished. Gabriello Fallopio, or Gabriel Fallopius
(1523–62),* was one of Vesalius' pupils and carried on in
the new tradition. He studied the reproductive system
and described the tubes leading from the ovary to the
uterus. These are still known as Fallopian tubes.

Another Italian anatomist, Bartolommeo Eustachio, or
Eustachius (c.1500–74) was an opponent of Vesalius
and an upholder of Galen, but he, too, looked at the
human body and described what he saw. He rediscovered
Alcmaeon's tube, running from ear to throat; this is now
known as the "Eustachian tube."

The refreshing new look in anatomy spread to other
branches of biology. The Hippocratic belief in the physician's light hand had, in later centuries, given way to
harsh remedies indeed. So crude did matters become, in
fact, that surgery, in early modern times, was not considered the concern of the physician but was left to the
barbering profession, which thus cut flesh as well as hair.
Perhaps because the barber-surgeons were weak on theory,
they relied heavily on drastic treatment. Gunshot wounds
were disinfected with boiling oil and bleeding was
stopped by charring the vessels shut with a red-hot iron.

The French surgeon, Ambroise Paré (1517–90), helped
change that. He began life as a barber's apprentice, joined
the army as a barber-surgeon, and introduced startling
innovations. He used gentle ointments (at room temperature) for gunshot wounds and stopped bleeding by
tying off the arteries. With an infinitesimal fraction of
the earlier pain, he effected far more frequent cures. He

---

* It was an age when Latin was the language of scholarship and
when many scholars used Latinized versions of their actual names.

is sometimes called, therefore, "the father of modern surgery."

Paré also devised clever artificial limbs, improved obstetrical methods, and wrote French summaries of the works of Vesalius so that other barber-surgeons, unlearned in Latin, might gather some facts concerning the structure of the human body, before hacking away at random.

And before long, just as the anatomists had to step down from the lecture platform and perform their own dissections, so physicians doffed their academic disdain and stooped to perform operations.

The Circulation of the Blood

Rather more subtle than the matter of the appearance and arrangement of the component parts of the body, which is the subject matter of anatomy, is the study of the normal functioning of those parts. The latter is *physiology*. The Greeks had made little progress in physiology and most of their conclusions were wrong. In particular, they were wrong about the functioning of the heart.

The heart is clearly a pump; it squirts blood. But where does the blood come from, and where does it go? The early Greek physicians made their first error in considering the veins to be the only blood vessels. The arteries are usually empty in corpses and so these were thought to be air-vessels. (The very word "artery" is from Greek words meaning "air duct.")

Herophilus, however, had shown that both arteries and veins carried blood. Both sets of blood vessels are joined with the heart and the matter would then have solved itself neatly if some connection between veins and arteries had been found at the ends away from the heart. However, the most careful anatomical investigation showed that both veins and arteries branched into finer and finer vessels until the branches grew so fine they were lost to sight. No connection between them could be found.

Galen, therefore, suggested that the blood moved from one set of vessels to the other by passing through the heart from the right half to the left. In order to allow the blood to pass through the heart, he maintained there must be tiny holes passing through the thick, muscular partition that divided the heart into a right and left half. These holes were never observed, but for seventeen centuries after Galen, physicians and anatomists assumed they were there. (For one thing, Galen had said so.)

The Italian anatomists of the new age began to suspect that this might not be so, without quite daring to come out in open rebellion. For instance, Hieronymus Fabrizzi, or Fabricius (1537–1619), discovered that the larger veins possessed valves. He described these accurately and showed how they worked. They were so arranged that blood could flow past them toward the heart without trouble. The blood, however, could not flow back away from the heart without being caught and trapped in the valves.

The simplest conclusion from this would be that the blood in the veins could travel in only one direction, toward the heart. This, however, interfered with Galen's notion of a back-and-forth motion and Fabricius only dared go as far as to suggest that the valves delayed (rather than stopped) the backward flow.

But Fabricius had a student, an Englishman named William Harvey (1578–1657), who was made of sterner stuff. After he returned to England, he studied the heart and noted (as had some anatomists before him) that there were one-way valves there, too. Blood could enter the heart from the veins, but valves prevented blood from moving back into the veins. Again, blood could leave the heart by way of the arteries but could not return to the heart because of another set of one-way valves. When Harvey tied off an artery, the side toward the heart bulged with blood; when he tied off a vein, the side away from the heart bulged.

Everything combined to show that the blood did not ebb and flow but moved in one direction perpetually. Blood flowed from the veins into the heart and from the heart into the arteries. It never backtracked.

Harvey calculated, furthermore, that in one hour the heart pumped out a quantity of blood that was three times the weight of a man. It seemed inconceivable that blood could be formed and broken down again at such a rate. Therefore, the blood in the arteries had to be returned to the veins someplace outside the heart, through connecting vessels too fine to see. (Such invisible vessels were no worse than Galen's invisible pores through the heart muscle.) Once such connecting vessels were assumed, then it was easy to see that the heart was pumping the same blood over and over again—veins/heart/arteries/veins/heart/arteries/veins/heart/arteries. . . . Thus it was not surprising it could pump three times the weight of a man in one hour.

In 1628, Harvey published this conclusion and the evidence backing it in a small book of only seventy-two pages. It was printed in Holland (and filled with typographical errors) under the title *De Motu Cordis et Sanguinus* ("On the Motions of the Heart and Blood"). For all its small size and miserable appearance, it was a revolutionary book that fitted the times perfectly.

Those were the decades when the Italian scientist, Galileo Galilei (1564–1642), was popularizing the experimental method in science and, in so doing, completely destroyed Aristotle's system of physics. Harvey's work represented the first major application of the new experimental science to biology and with it he destroyed Galen's system of physiology and established modern physiology. (Harvey's calculation of the quantity of blood pumped by the heart represented the first important application of mathematics to biology.)

The older school of physicians inveighed bitterly against Harvey, but nothing could be done against the facts. By the time of Harvey's old age, even though the connecting vessels between arteries and veins remained undiscovered, the fact of the circulation of the blood was accepted by biologists generally. Europe had thus stepped definitely and finally beyond the limits of Greek biology.

Harvey's new theory opened a battle between two opposing views of life, a battle that has filled the history of modern biology, and one that is not entirely settled even yet.

According to one major view of life, living things are considered essentially different from inanimate matter so that one cannot expect to learn the nature of life from studies on nonliving objects. In a nutshell, this is the view that there are two separate sets of natural law: one for living and one for nonliving things. This is the "vitalist" view.

On the other hand, one can view life as highly specialized but not fundamentally different from the less intricately organized systems of the inanimate universe. Given enough time and effort, studies of the inanimate universe will provide enough knowledge to lead to an understanding of the living organism itself, which, by this view, is but an incredibly complicated machine. This is the "mechanist" view.

Harvey's discovery was, of course, a blow in favor of the mechanist view. The heart could be viewed as a pump and the current of blood behaved as one would expect a current of inanimate fluid to behave. If this is so, where does one stop? Might not the rest of a living organism be merely a set of complicated and interlocking mechanical systems? The most important philosopher of the age, the Frenchman, René Descartes (1596–1650), was attracted by the notion of the body as a mechanical device.

In the case of man, at least, such a view was dangerously against the accepted beliefs of the day, and Descartes was careful to point out that the human body-machine did not include the mind and soul, but only the animal-like physical structure. With respect to mind and soul he was content to remain vitalist. Descartes made the suggestion that the interconnection between body and mind-soul was through a little scrap of tissue pendant from the brain, the "pineal gland." He was seduced into this belief by the mistaken feeling that only the human being possessed a pineal gland. This quickly proved not to be so. Indeed the pineal gland in certain primitive reptiles is far better developed than in the human.

Descartes' theories, though possibly wrong in details, were nevertheless very influential, and there were physiologists who attempted to hammer home the mechanist view in elaborate detail. Thus, the Italian physiologist, Giovanni Alfonso Borelli (1608–79), in a book appearing the year after his death, analyzed muscular action by treating muscle-bone combinations as a system of levers. This proved useful and the laws that held for levers made of wood held exactly for levers made of bone and muscle. Borelli tried to apply similar mechanical principles to other organs, such as the lungs and stomach, but there he was less successful.

The Beginnings of Biochemistry

Of course, the body may be viewed as a machine, without necessarily considering it merely a system of levers and gears. There are methods of performing tasks other than by the purely physical interlocking of components. There is chemical action, for instance. A hole might be punched in a piece of metal by means of a hammer and spike, but it might also be formed by the action of acid.

The first chemical experiments on living organisms were conducted by a Flemish alchemist, Jan Baptista van Helmont (1577–1644), who was Harvey's contemporary. Van Helmont grew a willow tree in a weighed quantity of soil and showed that after five years, during which time he added only water, the tree had gained 164 pounds, while the soil had lost only two ounces. From this, he deduced that the tree did not derive its substance primarily from the soil (which was right) and that it derived it instead from the water (which was wrong, at least in part). Van Helmont did not, unfortunately, take the air into account and this was ironical, for he was the first to study airlike substances. He invented the word "gas" and discovered a vapor which he called "*spiritus sylvestris*" ("spirit of the wood") which, as it later turned out, was the gas we call carbon dioxide which is, in fact, the major source of a plant's subsistence.

Van Helmont's first studies of the chemistry of living organisms (*biochemistry*, we now call it) began to develop and grow in the hands of others. An early enthusiast was Franz de la Boe (1614–72), usually known by his Latinized name, Franciscus Sylvius. He carried the concept to an extreme of considering the body a chemical device altogether. He felt that digestion was a chemical process, for instance, and that its workings were rather similar to the chemical changes that went on in fermentation. In this he turned out to be correct.

He also supposed that the health of the body depended upon the proper balance of its chemical components. In this, too, there are elements of truth, though the state of knowledge in Sylvius' time was far too primitive to make more than a beginning in this direction. All Sylvius could suggest was that disease was an expression of a superfluity or a deficiency of acid.

The Microscope

The great weakness in Harvey's theory of circulation was that he could not show that the arteries and veins ever actually met. He could only suppose that the connections existed but were too small to see. At the time of his death, the matter was still unsettled and might have remained so forever if mankind had been forced to rely on its unaided eyes. Fortunately it did not have to.

Even the ancients had known that curved mirrors and hollow glass spheres filled with water seemed to have a magnifying effect. In the opening decades of the seventeenth century men began to experiment with lenses in order to increase this magnification as far as possible. In this, they were inspired by the great success of that other lensed instrument, the telescope, first put to astronomical use by Galileo in 1609.

Gradually, enlarging instruments, or microscopes (from Greek words meaning "to view the small") came into use. For the first time, the science of biology was broadened and extended by a device that carried the human sense of vision beyond the limit that would otherwise be imposed upon it. It enabled naturalists to describe small creatures with a detail that would have been impossible without it, and it enabled anatomists to find structures that could not otherwise have been seen.

The Dutch naturalist, Jan Swammerdam (1637–80), spent his time observing insects under the microscope and producing beautiful drawings of the tiny details of their anatomy. He also discovered that blood was not a uniform red liquid, as it appeared to the eye, but that it contained numerous tiny bodies that lent it its color. (We now know those bodies as red blood corpuscles.) The English botanist, Nehemiah Grew (1641–1712), studied plants under the microscope and, in particular, their reproductive organs. He described the individual pollen grains they produced. A Dutch anatomist, Regnier de Graaf (1641–73), performed analogous work on animals. He studied the fine structure of the testicles and the ovaries. In particular, he described certain little structures of the ovary that are still called "Graafian follicles."

More dramatic than any of these discoveries was that of the Italian physiologist, Marcello Malpighi (1628–94). He, too, studied plants and insects, but among his early work was the study of the lungs of frogs. Here he found a complex network of blood vessels, too small to see individually, which were everywhere connected. Moreover, when he traced these small vessels back to their coalescence into larger vessels, the latter proved to be veins in one direction, arteries in the other.

Arteries and veins were, therefore, indeed connected by a network of vessels too small to be seen with the unaided eye, as Harvey had supposed. These microscopic vessels were named "capillaries" (from Latin words meaning "hairlike," though actually they are much finer than hairs). This discovery, first reported in 1660, three years after Harvey's death, completed the theory of the circulation of the blood.

Yet it was not Malpighi, either, who really put microscopy on the map, but a Dutch merchant, Anton van Leeuwenhoek (1632–1723), to whom microscopy was merely a hobby, but an all-absorbing one.

The early microscopists, including Malpighi, had used systems of lenses which, they rightly decided, could produce greater magnifications than a single lens alone could. However, the lenses they used were imperfect, possessing surface irregularities and inner flaws. If too much magnification was attempted, details grew fuzzy.

Van Leeuwenhoek, on the other hand, used single lenses, tiny enough to be built out of small pieces of flawless glass. He ground these with meticulous care to the point where he could get clear magnification of up to 200-fold. The lenses were, in some cases, no larger than the head of a pin, but they served Van Leeuwenhoek's purposes perfectly.

He looked at everything through his lenses and was able to describe red blood corpuscles and capillaries with greater detail and accuracy than the original discoverers, Swammerdam and Malpighi, could. Van Leeuwenhoek actually saw blood moving through the capillaries in the tail of a tadpole so that, in effect, he saw Harvey's theory in action. One of his assistants was the first to see the spermatozoa, the tiny tadpolelike bodies in male semen.

Most startling of all, though, was his discovery, when looking at stagnant ditch water under his lens, of the existence of tiny creatures, invisible to the naked eye, that, nevertheless, seemed to have all the attributes of life. These "animalcules" (as he called them) are now known as "protozoa" from Greek words meaning "first animals." Thus it became apparent that not only did objects exist too small to be seen by the naked eye, but *living* objects of that sort existed. A broad new biological territory thus opened up before the astonished gaze of men, and *microbiology* (the study of living organisms too small to be seen by the naked eye) was born.

In 1683, Van Leeuwenhoek even caught a fugitive glimpse of creatures considerably smaller than the protozoa. His descriptions are vague, of necessity, but it seems quite certain that his eye was the first in history to see what later came to be known as "bacteria."

The only other discovery of the era to match Van Leeuwenhoek's work, at least in future significance, was that of the English scientist, Robert Hooke (1635–1703). He was fascinated by microscopes and did some of the best of the early work. In 1665, he published a book, *Micrographia*, in which are to be found some of the most beautiful drawings of microscopic observations ever made. The most important single observation was that of a thin slice of cork. This, Hooke noted, was made up of a fine pattern of tiny rectangular chambers. He called these "cells," a common term for small rooms, and in later years, this discovery was to have great consequences.

Microscopy languished through the eighteenth century, chiefly because the instrument had reached the limit of its effectiveness. It was not till 1773, nearly a hundred years after Van Leeuwenhoek's original observation, that a Danish microbiologist, Otto Friderich Müller (1730–84), could see bacteria well enough to describe the shapes and forms of the various types.

One of the flaws of the early microscopes was that their lenses broke up white light into its constituent colors. Small objects were surrounded by rings of color ("chromatic aberration") that obscured fine detail. About 1820, however, "achromatic microscopes," which did not produce such rings of color, were devised. During the nineteenth century, therefore, the microscope was able to lead the way to new and startling areas of biologic advance.

CHAPTER 4

Classifying Life

Spontaneous Generation

The discoveries made by the microscope in the mid-seventeenth century seemed to blur the distinction between living and nonliving matter. It reopened a question that had seemed on the verge of a settlement. That question involved the origin of life or, at least, of the simpler forms of life.

While it was easy to see that human beings and the larger animals arose only from the bodies of their mothers, or from eggs laid by the mothers, this was not so clear in the case of smaller animals. It was taken for granted until modern times that creatures such as worms and insects grew out of decaying meat and other corruption. Such an origin of life from nonlife was referred to as "spontaneous generation."

The classic example presented as evidence for the existence of spontaneous generation was the appearance of maggots on decaying meat. It seemed obvious that

these small wormlike organisms had formed out of the
dead meat and almost all biologists accepted this fact.
One of the few exceptions, however, was Harvey who, in
his book on the circulation of the blood, speculated that
perhaps such small living things grew out of seeds or
eggs that were too small to be seen. (This was an easy
point for a biologist to make who was being forced to
postulate the existence of blood vessels too small to see.)

An Italian physician, Francesco Redi (1626–97), read
Harvey, was impressed, and decided to put the matter
to the test. In 1668, he prepared eight flasks with a variety
of kinds of meat inside. Four he sealed and four he left
open to the air. Flies could land only on the meat in
the open vessels and only the meat in those vessels bred
maggots. The meat in the sealed vessels decayed and
turned putrid but developed no maggots. Redi repeated
the experiment, covering some of the vessels with gauze,
rather than sealing them completely. In this way, air could
get at the meat freely, but flies would still be kept off.
Again, no maggots developed.

It seemed then that maggots developed not out of
meat but out of the eggs of flies. At this point, biological
thinking might well have veered off the concept of spontaneous generation altogether. However, the effect of
Redi's experiment was weakened by Van Leeuwenhoek's
contemporaneous discovery of protozoa. After all, flies
and maggots are still fairly complicated organisms, though
simple compared to men. Protozoa were themselves no
larger than flies' eggs, if as large, and were extremely simple living things. Surely, *they* could form by spontaneous
generation. The argument seemed upheld by the fact
that if nutritive extracts containing no protozoa were allowed to stand, the little creatures soon appeared in large
numbers. The matter of spontaneous generation became
part of the broader argument that was to reach new intensity in the eighteenth and nineteenth centuries: that
of the vitalists versus the mechanists.

The philosophy of vitalism was stated clearly by a German physician, Georg Ernst Stahl (1660-1734). Stahl is most famous for his theories concerning "phlogiston," a substance he supposed existed in substances that, like wood, could burn, or, like iron, could rust. When wood burned or iron rusted, phlogiston (Stahl said) was released into the air. To account for the fact that rusting metals gained weight, some chemists suggested that phlogiston had negative weight. When it was lost the metal therefore grew heavier. This theory proved very attractive to chemists and it was accepted by most of them throughout the eighteenth century.

However, in among Stahl's voluminous writings were also important views on physiology, particularly in a book on medicine which he published in 1707. He stated flatly that living organisms are not governed by physical laws but by laws of a completely different type. Little could be learned about biology, in his view, through the study of the chemistry and physics of the inanimate world. Opposed to him was the Dutch physician, Hermann Boerhaave (1668-1738), the most famous medical man of his times (sometimes called "the Dutch Hippocrates"). In his own book on medicine, he discusses the body in detail and tries to show how all its activity follows the laws of physics and chemistry—the mechanistic view.

For mechanists, who held that the same laws governed both the animate and inanimate worlds, microorganisms had a special importance. They seemed to serve almost as a bridge between life and nonlife. If it could be shown that such microorganisms actually formed from dead matter, the bridge would be complete—and easily crossed.

By the same token, the vitalist view, if valid, would require that, however simple life might be, there must still remain an unbridgeable gulf between it and inanimate matter. Spontaneous generation would not, by the strict vitalist view, be possible.

During the eighteenth century, however, mechanists

and vitalists did not line up solidly for and against (respectively) spontaneous generation, for religious views played a role, too. It was felt that the Bible described spontaneous generation in certain places so that many vitalists (who were generally the more conservative in religion) felt it necessary to back belief in the development of life from nonlife.

In 1748, for instance, an English naturalist, John Turberville Needham (1713–81), who was also a Catholic priest, brought mutton broth to a boil and placed it in a corked test tube. After a few days the broth was found to be swarming with microorganisms. Since Needham assumed that the initial heating had sterilized the broth, he concluded that the microorganisms had arisen out of dead material and that spontaneous generation, at least for microorganisms, had been proved.

One skeptic in this respect was the Italian biologist, Lazzaro Spallanzani (1729–99). He felt that the period of heating had been insufficiently prolonged and had not sterilized the broth in the first place. In 1768, therefore, he prepared a nutritive solution which he brought to a boil and then continued to boil for between one half and three quarters of an hour. Only then did he seal it in a flask and now microorganisms did *not* appear.

This seemed conclusive, but believers in spontaneous generation found a way out. They maintained that there was a "vital principle" in the air, something unperceived and unknown, which made it possible to introduce the capacity for life into inanimate matter. Spallanzani's boiling, they claimed, destroyed that vital principle. For nearly another century, then, the issue was to remain in doubt.

Arranging the Species

The argument over spontaneous generation was, in a sense, one over the problem of classifying life; whether to place it as eternally separate from nonlife or to allow a series of gradations. The seventeenth and eighteenth centuries saw, also, the development of attempts to classify the various forms present within the realm of life itself, and this was to serve as the start of an even more serious controversy than the one over spontaneous generation, a controversy that was to reach its climax in the nineteenth century.

To begin with, life forms can be divided into separate *species*, a word that is very difficult, actually, to define precisely. In a rough sense, a species is any group of living things that can mate freely among themselves and can, as a result, bring forth young like themselves which are also capable of mating freely to produce still another generation and so on. Thus, all human beings, whatever the superficial differences among them, are considered to belong to a single species because, as far as is known, men and women can breed freely among themselves regardless of those differences. On the other hand, the elephant of India and the elephant of Africa, although they look very much like the same sort of beast to the casual eye, are separate species, since a male of one group cannot mate and produce young with a female of the other.

Aristotle had listed five hundred species of animals, and Theophrastus as many species of plants. In the two thousand years since their time, however, continued observation had revealed more species and the broadening of the known world had unloosed a veritable flood of reports of new kinds of plants and animals that no ancient naturalist had ever seen. By 1700, tens of thousands of species of plants and animals had been described.

In any listing of even a limited number of species, it is very tempting to group similar species together. Almost anyone would naturally group the two species of elephants, for instance. To find a systematic method of grouping tens of thousands of species in a manner to suit biologists generally is no easy matter, and the first to make a major attempt in this direction was an English naturalist, John Ray (1628–1705).

Between 1686 and 1704, he published a three-volume encyclopedia of plant life in which he described 18,600 species. In 1693, he prepared a book on animal life that was less extensive but in which he attempted to make a logical classification of the different species into groups. He based the groups largely on the toes and teeth.

For instance, he divided mammals into two large groups: those with toes and those with hoofs. He divided the hoofed animals into one-hoofed (horses), two-hoofed (cattle, etc.), and three-hoofed (rhinoceros). The two-hoofed mammals he again divided into three groups: those which chewed the cud and had permanent horns (goats, etc.), those which chewed the cud and had horns that were shed annually (deer), and those which did not chew the cud (swine.)

Ray's system of classification was not kept, but it had the interesting feature of dividing and subdividing, and this was to be developed further by the Swedish naturalist, Carl von Linné (1707–78), usually known by the Latinized name, Carolus Linnaeus. By his time, the number of known species of living organisms stood at a minimum of 70,000; and Linnaeus, in 1732, traveled 4600 miles hither and yon through northern Scandinavia (certainly not a lush habitat for life) and discovered a hundred new species of plants in a short time.

While still in college, Linnaeus had studied the sexual organs of plants, noted the manner in which they differed from species to species, and decided to try to form a system of classification based on this. The project grew broader with time and in 1735, he published *System Naturae*, in which he established the system of classifying species which is the direct ancestor of the system used today. Linnaeus is therefore considered the founder of *taxonomy*, the study of the classification of species of living things.

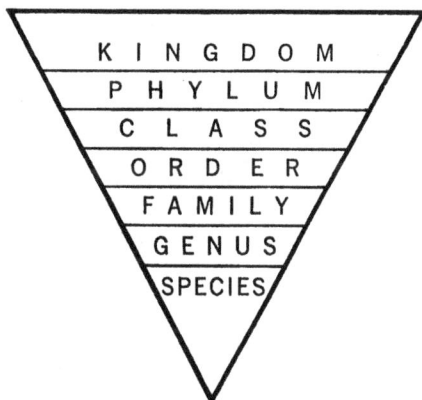

FIGURE 1. Diagram showing, in descending order, the main classifications—from Kingdom to Species—into which living things are placed by taxonomists.

Linnaeus systematically grouped similar species into "genera" (singular, "genus," from a Greek word meaning "race"). Similar genera were grouped into "orders," and similar orders into "classes." All the known animal species were grouped into six classes: mammals, birds, reptiles, fishes, insects and "vermes." These major divisions were not, actually, as good as those of Aristotle two thousand years before, but the systematic division and subdivision made up for that. The shortcomings were patched up easily enough later on.

To each species, Linnaeus gave a double name in Latin; first the genus to which it belonged, then the specific name. This form of "binomial nomenclature" has been retained ever since and it has given the biologist an international language for life forms that has eliminated incalculable amounts of confusion. Linnaeus even supplied the human species with an official name; one that it has retained ever since—*Homo sapiens.*

Approach to Evolution

Linnaeus' classification, beginning with extremely broad groups and dividing into successively narrower groups, seemed like a literal "tree of life." Looking upon the representation of such a tree, however diagrammatic, it was almost inevitable that one would wonder whether the arrangement could be entirely accidental. Might not two closely related species have developed from a common ancestor, and might not two closely related ancestors have developed from a still more ancient and primitive ancestor? In short, might not the structure designed by Linnaeus have grown over the ages somewhat as a real tree might have grown? It was over this possibility that the greatest controversy in the history of biology arose.

To Linnaeus himself, a pious man devoted to the literal word of the Bible, this possibility was anathema. He insisted that every species had been separately created and that each had been maintained by divine Providence so that no species had been allowed to become extinct. His own system of classification reflected this belief, for it was based on external appearance and made no attempt to mirror possible relationships. (It was as though one were to group donkeys, rabbits, and bats into one category because all had long ears.) To be sure, if there were no relationships among species, it didn't matter how you grouped them; all arrangements were equally artificial and one might as well choose the most convenient.

Nevertheless, Linnaeus could not stop others from suggesting or supposing some process of "evolution" (the word itself did not become popular till the mid-nineteenth century) in which one species *did* develop from another, and in which there were natural relationships among species that ought to be reflected in the system of classification used. (In later life, even Linnaeus himself began to weaken and to suggest that new species might arise through hybridization.)

Even the French naturalist, Georges Louis Leclerc, Comte de Buffon (1707–88), easygoing, conservative, and cautious (he had collaborated with Needham in the latter's experiment on spontaneous generation, see page 34), could not help but dare the prevailing orthodoxy by suggesting such a thing.

De Buffon wrote a forty-four volume encyclopedia on natural history, as popular in his time as ever Pliny's had been, and as heterogeneous (but far more accurate). In it, he pointed out that some creatures had parts that were useless to them ("vestiges"), like the two shriveled toes a pig possessed on the sides of its two useful hoofs. Might they not represent toes that had once been full-sized and useful but that had shriveled with time? Might not whole organisms do the same? Might not an ape be a degenerated man, or a donkey a degenerated horse?

An English physician, Erasmus Darwin (1731–1802), wrote long poems dealing with botany and zoology in which he accepted the Linnaean system. In them he also adopted the possibility of changes in species brought on by environmental effects. (However, these views would undoubtedly be forgotten today, were it not for the fact that Erasmus Darwin was the grandfather of Charles Darwin, with whom evolutionary theory reached its climax.)

The coming of the French Revolution, the year after De Buffon's death, shook Europe to its core. An era of change was introduced in which old values were shattered, never again to be restored. The easy acceptance of King and Church as ultimate authorities vanished in one nation after another and it became possible to suggest scientific theories that would have been dangerous heresies earlier. Thus, De Buffon's views of the world of life were such as to make it unnecessary to deal very extensively with evolutionary doctrine. Some decades later, however, another French naturalist, Jean Baptiste de Monet, Chevalier de Lamarck (1744–1829), found it desirable to consider evolution in considerable detail.

Lamarck grouped the first four Linnaean classes (mammals, birds, reptiles, fish) as "vertebrates," animals possessing an internal vertebral column, or backbone. The other two classes (insects and worms) Lamarck named "invertebrates." (Although this twofold classification was quickly superseded, it remains in popular use among laymen.) Lamarck recognized the classes of insects and worms to be heterogeneous grab bags. He labored over them and reduced them to better order; raising them, indeed, to the level at which they stood in Aristotle's classification and beyond. He recognized, for instance, that the eight-legged spiders could not be classified with the six-legged insects, and that lobsters could not be lumped with starfish.

Between 1815 and 1822, Lamarck finally produced a gigantic seven-volume work entitled *Natural History of Invertebrates*, which founded modern invertebrate zoology. This work had already caused him to think about the possibility of evolution and he had published his thinking on the subject as early as 1801 and then, in greater detail, in 1809 in a book called *Zoological Philosophy*. Lamarck suggested that organs grew in size of efficiency if used much during life, and degenerated if not used; and that this growth or degeneration could then be passed on to the offspring. (This is often referred to as "inheritance of acquired characteristics.")

He used the then recently discovered giraffe as an example of what he meant. A primitive antelope, fond of browsing on the leaves of trees, would stretch its neck upward with all its might to get all the leaves it could. Tongue and legs would stretch, too. All these body parts would literally grow slightly longer as a result, and this lengthening, Lamarck suggested, would be passed on to the next generation. The new generation would start with longer parts and stretch them still further. Little by little, the antelope would turn into a giraffe.

The theory did not stand up, for there was no good evidence that acquired characteristics could be inherited. In fact, all the evidence that could be gathered indicated that acquired characteristics were *not* inherited. Even if such characteristics could be inherited, that might do for those which could undergo a voluntary stress as in the case of a stretched neck. But what about the giraffe's blotched skin which served as protective camouflage? How did that develop from an antelope's unblotched hide? Could the ancestral giraffe conceivably have tried to become blotched?

Lamarck died poor and neglected, and his theory of evolution was shrugged off. But it had opened the floodgates just the same. Evolution might have suffered a defeat but the mere fact that it had entered the battleground was significant. There would be other chances to fight later.

The Geological Background

A major difficulty that stood in the way of all theories of evolution was the apparent slowness of species change. In the memory of mankind there were no cases of one species turning into another. If such a process did take place, therefore, it must be exceedingly slow, requiring, perhaps, hundreds of thousands of years. Yet throughout medieval and early modern times, European scholars accepted the literal words of the Bible and considered the earth to be only some six thousand years old, and that left no time for an evolutionary process.

In 1785, came a change. James Hutton (1726–97), a Scottish physician who had taken up geology as a hobby, published a book called *Theory of the Earth*. In it, he reviewed the manner in which the action of water, wind and weather slowly changed the surface of the earth. He maintained that these actions had always proceeded in the same way and at the same rate ("the uniformitarian principle"). He then pointed out that to account for such gigantic changes as the building of mountains, the gouging out of river canyons and so on, vast ages of time were required. The earth, therefore, must be many millions of years old.

This new concept of the age of the earth was at first greeted with a most hostile reception, but it had to be admitted that it helped make sense of the fossils that were now beginning to preoccupy biologists. The word "fossil" comes from a Latin word meaning "to dig" and was originally applied to anything dug up out of the earth. However, the dug-up materials that excited most curiosity were stony objects that seemed to possess structures like those of living organisms.

It seemed quite unlikely that stones should mimic life forms by accident, so most scholars felt that they had to be once-living things that had somehow turned to stone. Many suggested they were remains of creatures destroyed by Noah's flood. If, however, the earth were as old as Hutton suggested, they might be extremely ancient remains that had very slowly replaced their ordinary substance by the stony material in the soil about them.

A new look at fossils came with William Smith (1769–1839), an English surveyor turned geologist. He surveyed routes for canals (then being built everywhere) and had the opportunity to observe excavations. He noted the manner in which rocks of different types and forms were arranged in parallel layers or "strata." He noted in addition that each stratum had its own characteristic form of fossil remains, not found in other strata. No matter how a stratum was bent and crumpled, even when it sank out of view and cropped up again miles away, it retained its characteristic fossils. Eventually, Smith was able to identify different strata by their fossil content alone.

If Hutton's views were correct, then it was reasonable

to suppose that the strata lay in the order in which they were very slowly formed, and that the deeper a particular stratum lay, the older it was. If the fossils were, indeed, the remains of once-living creatures, then the order in which they lived might be determined by the order of the strata in which they were to be found.

Fossils attracted the particular attention of a French biologist, Georges Léopold Cuvier (1769–1832). Cuvier studied the anatomy of different creatures, comparing them carefully, and systematically noting all similarities and differences, thus founding *comparative anatomy*. These studies made it possible for Cuvier to learn the necessary relationship of one part of a body with another so well that from the existence of some bones, he could infer the shape of others, the type of muscles that must be attached, and so on. In the end, he could reconstruct a reasonable approximation of the entire animal from a small number of parts.

It seems natural that a comparative anatomist should be interested in the classification of species. Cuvier extended Linnaeus' system by grouping the latter's classes into still larger groups. One he called "vertebrata" as Lamarck had done. Cuvier did not, however, lump the rest as invertebrates. Instead, he divided them into three groups: articulata (shelled animals with joints, such as insects and crustacea), mollusca (shelled animals without joints, such as clams and snails), and radiata (everything else).

These largest groups he called "phyla" (singular, "phylum," from a Greek word meaning "tribe"). Since Cuvier's day, the phyla have been multiplied until now some three dozen phyla of living creatures, both plant and animal, are recognized. In particular, the phylum of vertebrates has been extended to include some primitive animals without vertebral columns and it is now called "chordata."

Again because of his interest in comparative anatomy,

Cuvier based his own system of classification on those characteristics which indicated relationships of structure and functioning, rather than on the superficial similarities that guided Linnaeus. Cuvier applied his system of classification primarily to animals, but in 1810, the Swiss botanist, Augustin Pyramus de Candolle (1778–1841), applied it to plants as well.

Cuvier could not help but extend his system of classification to the fossils. To his experienced eye, which could build whole organisms out of parts, fossils did not merely resemble living things; they possessed features that placed them clearly in one or another of the phyla he had established. He could even classify them among the subgroups of the particular phylum to which they belonged. Thus, Cuvier pushed biological knowledge into the far past and established the science of *paleontology*, the study of ancient forms of life.

The fossils, as seen by Cuvier, seemed to represent the record of an evolution of species. The deeper and older a fossil was, the more it differed from existing life forms, and some could be placed in consecutive order in a manner that seemed to demonstrate gradual change.

Cuvier, however, was a pious man who could not accept the possibility of evolutionary changes. He adopted instead an alternative view that although the earth was indeed ancient, it underwent periodic catastrophes during which all life was wiped out. After each such catastrophe, new forms of life would appear, forms that were quite different from those that had previously existed. Modern forms of life (including man) were created after the most recent catastrophe. In this view, evolutionary processes were not needed to explain the fossils, and the biblical story, supposed to apply only to events after the last catastrophe, could be preserved.

Cuvier felt that four catastrophes were needed to explain the known distribution of fossils. However, as more and more fossils were discovered, matters grew more complicated and some of Cuvier's followers eventually postulated as many as twenty-seven catastrophes.

Such "catastrophism" was not in accord with the uniformitarianism of Hutton. In 1830, the Scottish geologist, Charles Lyell, began the publication of a three-volume book, *Principle of Geology*, in which he popularized Hutton's views and marshaled the evidence indicating that earth underwent only gradual and noncatastrophic changes. And, to be sure, continuing studies of fossils backed Lyell. There seemed no points at all in the records of the strata where *all* life was wiped out. Some forms survived each period where a catastrophe was suggested. Indeed, some forms now alive have existed virtually unchanged for many millions of years.

Catastrophism held out for a while among Cuvier's followers, particularly in France, but after Lyell's book appeared, it was clearly a dying belief. Catastrophism was the last scientific stand against the theory of evolution, and when it collapsed, some form of evolutionary concept simply had to be formulated. By the mid-nineteenth century, conditions were ripe—more than ripe—for such a development and the man to bring it about was on the scene.

CHAPTER 5

Compounds and Cells

Gases and Life

While species were being successfully classified, the science of life was being extended in a new and extremely fruitful direction. The study of chemistry was being revolutionized and chemists began to apply their techniques to living organisms as well as to inanimate systems. That this was a legitimate thing to do was clearly demonstrated in one early experiment on digestion.

Digestion is one function of the animal body that is relatively open to investigation. It does not take place within the body tissues themselves, but in the food canal which is open to the outside world and can be reached by way of the mouth. In the seventeenth century there had been a serious question as to whether digestion was a physical process involving the grinding action of the stomach, as suggested by Borelli (see page 26), or a chemical process involving the fermenting action of stomach juices, as suggested by Sylvius (see page 27).

A French physicist, René Antoine Ferchault de Réaumur (1683–1757), thought of a way of testing this. In 1752, he placed meat in small metal cylinders open at both ends (the ends being covered by wire gauze) and persuaded a hawk to swallow them. The metal cylinder protected the meat from any grinding action, while the wire gauze permitted stomach juices to enter, without allowing the meat to fall out. Hawks generally regurgitate indigestible matter and when Réaumur's hawk regurgitated the cylinder, the meat inside was found to be partially dissolved.

Réaumur double-checked by having the hawk swallow and regurgitate a sponge. The stomach juices that saturated the sponge were then squeezed out and mixed with meat. The meat slowly dissolved, and the issue was settled. Digestion was a chemical process and the role of chemistry in life was effectively dramatized.

In the eighteenth century, the study of gases, begun by Van Helmont (see page 27), was progressing with particular rapidity and becoming a glamorous field of study. It was inevitable that the connection of various gases with life be explored. An English botanist and chemist, Stephen Hales (1677–1761), was one of the explorers. He published a book in 1727, in which he described experiments by which he measured the rates of plant growth, and the pressure of sap, so that he is considered the founder of *plant physiology*. He also, however, experimented with a variety of gases and was the first to recognize that one of them, carbon dioxide, contributed somehow to the nourishment of plants. In this he corrected (or, rather, extended) Van Helmont's view that it was water alone out of which plant tissues were formed.

The next step was taken by the English chemist, Joseph Priestley (1733–1804) a half-century later. In 1774, he discovered the gas we now call oxygen. He found that it was pleasant to breathe and that mice were particularly frisky when placed in a bell jar containing oxygen. He further recognized the fact that plants increased the quantity of oxygen in the air. A Dutch physician, Jan Ingenhousz (1730–99), showed, moreover, that the process by which plants consumed carbon dioxide and produced oxygen took place only in the presence of light.

The greatest chemist of the age was the Frenchman, Antoine Laurent Lavoisier (1743–94). He emphasized the importance of accurate measurement in chemistry and used it to develop a theory of combustion that has been accepted as true ever since. According to this theory, combustion is the result of a chemical union of the burning material with the oxygen of the air. He showed also that, in addition to oxygen, air contains nitrogen, a gas that does not support combustion.

Lavoisier's "new chemistry" had its applications to life forms, too, for in some ways what applied to a candle applied to a mouse as well. When a candle is set to burning in a closed bell jar, oxygen is consumed and carbon dioxide is produced. The latter comes about through the combination of the carbon contained in the substance of the candle with the oxygen. When all or almost all the oxygen in the air within the bell jar is consumed, the candle goes out and will no longer burn.

The situation is similar for animal life. A mouse under a bell jar consumes oxygen and forms carbon dioxide; the latter through the combination of the carbon in its tissue substance with oxygen. As the oxygen level in the air drops, the mouse suffocates and dies. From the over-all point of view, plants consume carbon dioxide and produce oxygen, and animals consume oxygen and produce carbon dioxide. Plants and animals together, then, help maintain the chemical balance so that, in the long run, the atmospheric content of oxygen (21 per cent) and of carbon dioxide (0.03 per cent) remain steady.

Since a candle and an animal both produce carbon dioxide and consume oxygen, it seemed reasonable to Lavoisier to suppose that respiration was a form of combustion and that when a particular amount of oxygen was consumed, a corresponding quantity of heat was produced whether it was a candle or a mouse that was involved. His experiments in this direction were necessarily crude (considering the measuring techniques then available) and his results only approximate, but they seemed to bear out his contention.

This was a powerful stroke on the side of the mechanistic view of life, for it seemed to imply that the same chemical process was taking place in both living and nonliving matter. This made it that much more reasonable to suppose that the same laws of nature governed both realms as the mechanists insisted.

Lavoisier's point was strengthened as the science of physics developed during the first half of the nineteenth century. In those decades, heat was being investigated by a number of scientists whose interest was aroused by the growing importance of the steam engine. Heat, by means of the steam engine, could be made to do work, and so could other phenomena, such as falling bodies, flowing water, air in motion, light, electricity, magnetism, and so on. In 1807, the English physician, Thomas Young (1773–1829), suggested "energy" as a word to represent all phenomena out of which work could be obtained. It comes from Greek words meaning "work within."

The physicists of the early nineteenth century studied the manner in which one form of energy could be converted to another, and made increasingly refined measurements of such changes. By the 1840s, at least three men, an Englishman, James Prescott Joule (1818–89), and two Germans, Julius Robert von Mayer (1814–78) and Hermann Ludwig Ferdinand von Helmholtz (1821–94), had advanced the concept of the "conservation of energy." According to this concept, one form of energy might be freely converted into another, but the total amount of energy could neither be decreased nor increased in the process.

It seemed natural for such a broadly general law, based on a wide variety of meticulous measurements, to apply to living processes as well as nonliving. The mere fact that no living animal could continue living without obtaining energy continuously from its food made it seem that life processes could not create energy out of nothing. Plants did not eat and breathe in quite the same way animals did, but, on the other hand, they could not live unless they were periodically bathed in the energy of light.

Mayer, indeed, specifically stated that the source of all the various forms of energy on earth was the radiation of light and heat from the sun; and that this was likewise the source of the energy that powered living organisms. It was the direct energy source for plants and, through plants, for animals (including, of course, man).

The suspicion grew, then (and was to be amply demonstrated in the second half of the nineteenth century), that the law of conservation of energy applied as strictly to animate nature as to inanimate nature and that in this very important respect, life was mechanistic.

Organic Compounds

The vitalist position remained strong, however. Even if it became necessary to concede that the law of conservation of energy held for living systems as well as non-living; or that both bonfires and living animals consumed oxygen and produced carbon dioxide in similar fashion, these represented merely over-all limitations—like saying that both human beings and mountain tops were composed of matter. There still remained the vast question of detail within that limitation.

Might it not be that living organisms, though composed of matter, were made up of forms of matter not quite like that of the inanimate world, for instance? This question almost seemed to answer itself, in the affirmative.

Those substances that abounded in the soil, sea, and air were solid, stable, unchanging. Water, if heated, boiled and became vapor, but could be cooled to liquid water again. Iron or salt might be melted, but could be frozen once more to the original. On the other hand, substances obtained from living organisms—sugar, paper, olive oil—seemed to share the delicacy and fragility of the life forms from which they were derived. If heated, they smoked, charred, or burst into flame, and the changes they underwent were irreversible; the smoke and ash of burning paper did not become paper again upon cooling. Surely, then, it might be fair to suppose that two distinct varieties of matter were being dealt with here.

The Swedish chemist, Jöns Jakob Berzelius (1779–1848), suggested, in 1807, that substances obtained from living (or once-living) organisms be called "organic substances," while all others be referred to as "inorganic substances." He felt that while it was possible to convert organic substances to inorganic ones easily enough, the reverse was impossible except through the agency of life.

To prepare organic substances from inorganic, some vital force present only in living tissue had to be involved.

This view, however, did not endure for long. In 1828, a German chemist, Friedrich Wöhler (1800–82), was investigating cyanides and related compounds; compounds which were then accepted as inorganic. He was heating ammonium cyanate and found, to his amazement, that he obtained crystals that, on testing, proved to be urea. Urea was the chief solid constituent of mammalian urine and was definitely an organic compound.

Wöhler's discovery encouraged other chemists to tackle the problem of synthesizing organic substances out of inorganic ones, and success followed rapidly. With the work of the French chemist, Pierre Eugène Marcelin Berthelot (1827–1907), there remained no question that the supposed wall between inorganic and organic had broken down completely. In the 1850s, Berthelot synthesized a number of well-known organic compounds, such as methyl alcohol, ethyl alcohol, methane, benzene, and acetylene from compounds that were clearly inorganic.

With the development of appropriate analytical techniques in the first decades of the nineteenth century, chemists found that organic compounds were made up chiefly of carbon, hydrogen, oxygen, and nitrogen. Before long they learned to put these substances together in such a way that the resulting compound had the general properties of organic substances but did not actually occur in living creatures.

The latter half of the nineteenth century saw myriads of "synthetic organic compounds" formed, and it was no longer possible to define organic chemistry as being the study of compounds produced by life forms. To be sure, it was still convenient to divide chemistry into two parts, organic and inorganic, but these came to be defined as "the chemistry of carbon compounds" and "the chemistry of compounds not containing carbon," respectively. Life had nothing to do with it.

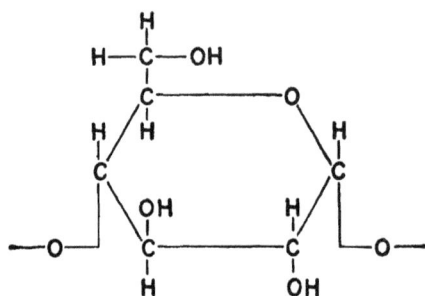

CARBOHYDRATE

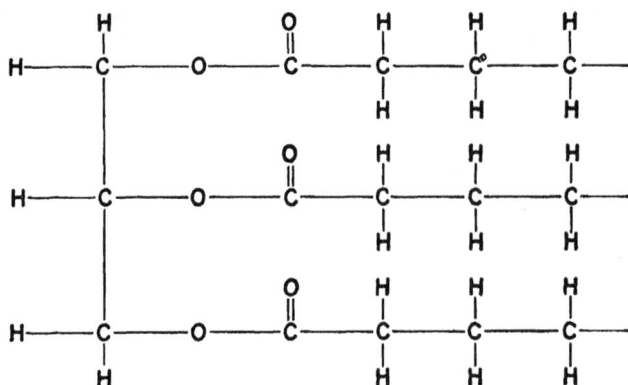

FAT (Lipid)

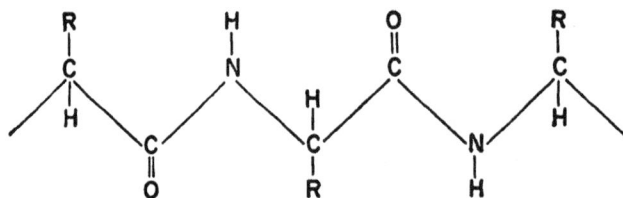

PROTEIN

FIGURE 2. The chemical formulas for the three classes of organic substances of which all things are composed: carbohydrate, lipid (fat), and protein. The carbohydrate is a chain of
six-carbon sugar units, only one unit of which is shown. The fat
in this illustration is palmitin, one of the commonest, and consists of the glycerol atoms at the left and a long chain of fatty

And yet there was considerable room for the vitalists to retreat. The organic compounds formed by nineteenth-century chemists were relatively simple ones. There existed in living tissue many substances so complex that no nineteenth-century chemist could hope to duplicate them.

These more complex compounds fell into three general groups, as the English physician, William Prout (1785–1850), was the first to state, specifically, in 1827. The names we now give the groups are "carbohydrates," "lipids," and "proteins." The carbohydrates (which include sugars, starch, cellulose, and so on) are made up of carbon, hydrogen and oxygen only, as are the lipids (which include fats and oils). The carbohydrates, however, are relatively rich in oxygen, while the lipids are poor in it. Again, the carbohydrates are either soluble in water to begin with or are easily made soluble by the action of acids, whereas the lipids are insoluble in water.

The proteins, however, were the most complex of these three groups, the most fragile, and, seemingly, the most characteristic of life. Proteins contained nitrogen and sulfur as well as carbon, hydrogen, and oxygen, and, though usually soluble in water, coagulated and became insoluble when gently heated. They were at first called "albuminous substances," because a good example was to be found in egg white which, in Latin, is called "albumen." In 1838, however, a Dutch chemist, Gerard Johann Mulder (1802–80), recognizing the importance of the albuminous substances, coined the word "protein" from Greek words meaning "of first importance."

Throughout the nineteenth century, the vitalists pinned their hopes, not on organic substances generally, but on the protein molecule.

The developing knowledge of organic chemistry also

acids (partially shown at the right). The protein formula illustrated here is a portion of a polypeptide chain, the backbone of a protein molecule. The letter R represents the side chains of amino acids (see Figure 6 for detail). (After a drawing in *Scientific American*.)

contributed to the evolutionary concept. All species of living things were composed of the same classes of organic substances: carbohydrates, lipids, and proteins. To be sure, these differed from species to species but the differences were minor. Thus, a palm tree and a cow are extremely different creatures, but the fat produced from coconuts and from milk are different in only trivial ways.

Furthermore, it became clear to chemists of the mid-nineteenth century that the complicated structure of carbohydrates, lipids, and proteins could be broken down to relatively simple "building blocks" in the course of digestion. The building blocks were identical for all species and only the details of combining them seemed different. One creature could feed upon another widely different one (as when a man eats a lobster or a cow eats grass) because the complex substances of the food are broken down to the building blocks held in common by eater and eaten; and these building blocks are absorbed and then built up again into the complex substances of the creature who feeds.

From the chemical standpoint, then, it would seem that all life, however various in outer appearance, is one. If this is so, then evolutionary changes of one species to another would seem to be mere matters of detail; and to require no truly fundamental shift. This view increased the plausibility of the evolutionary concept even if, in itself, it did not establish that concept.

Tissues and Embryos

Nor did the biologist have to depend on the somewhat alien world and work of the chemist to become aware of the basic unity of life. The developing excellence of the microscope eventually made this point visible to the eye.

At first, the microscope made too much visible to the eye, or, rather, to the imagination. Some of the early microscopists, fascinated by the glimpse into the infinitesimal, insisted on making out details beyond the power of their poor instruments to offer them. Thus, they painstakingly drew pictures of microscopic human figures ("homunculi") within the spermatozoa of the semen they studied.

They imagined, too, there might be no end to smallness. If an egg or sperm already contained a tiny figure, that tiny figure might contain within it a still tinier one that was someday to be its offspring and so on indefinitely. Some even tried to calculate how many homunculi within homunculi within homunculi might have existed in Eve in the first place; and wondered whether the human race might not come to an end when those nested generations were exhausted. This was the doctrine of "preformation" and was clearly an antievolutionary view since, according to it, all possible members of a species already existed in the first member of that species and there was no reason to suppose that there would be a change of species anywhere along the line.

The first major attack on this point of view came from a German physiologist, Caspar Friedrich Wolff (1733–94). In a book published in 1759, when he was only 26, he described his observations of the development of growing plants. He noted that the tip of a growing-plant shoot consisted of undifferentiated and generalized structures. As the tip grew, it specialized, however, and one bit eventually developed into a flower while another bit (completely indistinguishable at first) developed into a leaf. Later, he extended his observations to animals such as the embryonic chick. Undifferentiated tissue, he showed, gave rise to the different abdominal organs through gradual specialization. This was the doctrine of "epigenesis," an expression first used by William Harvey in a book on the birth of animals, published in 1651.

From this viewpoint, all creatures, however different in appearance, developed out of simple blobs of living mat-

ter and were alike in their origins. Living things did *not*
develop out of a tiny, but already specialized, organ or
organism.

Even fully developed organisms were not as different
as they might appear to be, when studied properly. A
French physician, Marie François Xavier Bichat (1771–
1802), working without a microscope (!) was able to
show, in the last years of his short life, that various organs
consisted of several components of different appearance.
These components he named "tissues" and thus founded
the science of *histology*, the study of tissues. It turned
out there were not very many different tissues (some important varieties in animals are epithelial, connective,
muscle and nerve tissues) and that different organs of
different species were built up out of these few varieties.
Particular tissues did not differ from species to species
as radically as the whole organisms did.

And one can go still further than that, too. As was explained earlier in the book (see page 30), Hooke, in the
mid-seventeenth century, had observed that cork was divided up into tiny rectangular chambers which he called
cells. These were empty, but then cork was a dead tissue.
Later investigators, studying living, or recently living, tissues under the microscope came to realize that these, too,
were built up out of tiny, walled-off units.

In living tissue, the units are not empty, but are filled
with a gelatinous fluid. This fluid was eventually to receive a name thanks to a Czech physiologist, Johannes
Evangelista Purkinje (1787–1869). In 1839, he referred
to living embryonic material within an egg as "protoplasm," from Greek words meaning "first formed." The
German botanist, Hugo von Mohl (1805–72), adopted
the term the next year but applied it to the material
within tissues generally. Although the partitioned units of
living tissue were not empty, Hooke's word "cell" continued to be applied to them.

Cells were more and more commonly found and a number of biologists speculated that they might exist universally within living tissue. This belief crystallized in 1838, when a German botanist, Matthias Jakob Schleiden (1804–81), maintained that all plants were built up of cells and that it was the cell that was the unit of life; a little living thing out of which entire organisms were built.

In the next year, a German physiologist, Theodor Schwann (1810–82), extended and amplified this idea. He pointed out that all animals, as well as all plants, were built up out of cells; that each cell was surrounded by a membrane separating it from the rest of the world; and that Bichat's tissues were built up of cells of a particular variety. Usually, then, Schleiden and Schwann share the credit for the "cell theory," though many others also contributed, and with them begins the science of *cytology* (the study of cells).

The assumption that the cell was the unit of life would be particularly impressive if it could be shown that a cell was capable of independent life, that, to be living, it was not necessary for it to be combined into conglomerates of billions and trillions. That some cells actually were capable of independent life was shown by a German zoologist, Karl Theodor Ernst von Siebold (1804–85).

In 1845, Siebold published a book on comparative anatomy which dealt in detail with protozoa, the little animals first detected by Van Leeuwenhoek (see page 30). Siebold made it quite clear that protozoa had to be considered as consisting of single cells. Each protozoon was surrounded by a single membrane and possessed within itself all the essential faculties of life. It ingested food, digested it, assimilated it, and discarded wastes. It sensed its environment and responded accordingly. It grew, and it reproduced by dividing in two. To be sure, the protozoa were generally larger and more complex than the cells making up a multicellular organism such as man; but then the protozoan cell had to be, for it retained all necessary abilities that made independent life possible, whereas individual cells of a multicellular creature could afford to discard much of this.

Even multicellular organisms could be used to demonstrate the importance of individual cells. The Russian biologist, Karl Ernst von Baer (1792–1876), had, in 1827, discovered the mammalian egg within the Graafian follicle (see page 28) and then went on to study the manner in which the egg developed into an independently living creature.

Over the course of the next decade, he published a large two-volume textbook on the subject, thus founding the science of *embryology* (the study of the embryo, or developing egg). He revived Wolff's theory of epigenesis (which had been largely ignored in its own time) in more detailed and better-substantiated form, showing that the developing egg forms several layers of tissue, each of which is undifferentiated to begin with, but out of each of which various specialized organs developed. These original layers he called "germ layers" ("germ" being a general term for any small object containing the seed of life).

The number of such germ layers was finally fixed at three, and in 1845, the German physician, Robert Remak (1815–65), gave them the names by which they are still known. These are "ectoderm" (from Greek words meaning "outer skin"), "mesoderm" ("middle skin"), and "endoderm" ("inner skin").

The Swiss physiologist, Rudolf Albert von Kölliker (1817–1905), pointed out, in the 1840s, that the egg and sperm were individual cells. (Later, the German zoologist, Karl Gegenbaur [1826–1903], went on to demonstrate that even the large eggs of birds were single cells.) The fusion of sperm and egg formed a "fertilized ovum" which, Kölliker showed, was still a single cell. (This fusion, or "fertilization," initiated the development of the embryo. Although biologists were already assuming, by mid-nineteenth century, that the process took place, and though a number of observations supporting this assumption were made over the preceding decades, it was not actually described in detail until 1879, when the Swiss zoologist, Hermann Fol, witnessed the fertilization of a starfish egg by a sperm.)

By 1861, Kölliker had published a textbook on embryology in which Baer's work was reinterpreted in terms of the cell theory. Every multicellular organism began as a single cell, the fertilized ovum. As the fertilized ovum divided and redivided, the resulting cells were not very different to begin with. Slowly, however, they specialized in different directions until all the complexly interrelated structures of the adult form were produced. It was epigenesis reduced to cellular terms.

The concept of the unity of life was greatly strengthened. One could scarcely differentiate between the fertilized ovum of a man, a giraffe, and a mackerel and, as the embryo developed, differences were produced only gradually. Small structures in the embryos, scarcely distinguishable at first, might develop into a wing in one case, an arm in another, a paw in a third, and a flipper in still a fourth. Baer felt, quite strongly, that relationships among animals could more properly be deduced by comparing embryos than by comparing adult structures, so that he is also the founder of *comparative embryology*.

The change from species to species, viewed through the process of cellular development, seemed a matter of detail only, and to be well within the capacity of some evolutionary process to bring about. Baer was able to show, for instance, that the early vertebrate embryo possessed a "notochord" temporarily. This is a stiff rod running the length of the back and there are very primitive fishlike creatures that possess such a structure throughout life. These primitive creatures were first studied and described in the 1860s by the Russian zoologist, Alexander Kowalewski (1840–1901).

In vertebrates, the notochord is quickly replaced by a

spinal cord of jointed vertebrae. Nevertheless, even the temporary appearance of the notochord seems to show a relationship to the animals described by Kowalewski. It is for this reason that the vertebrates and these few invertebrates are lumped together in the phylum, Chordata. Moreover, it is even attractive to suppose that the notochord, appearing so briefly in the vertebrate embryo (even in the human embryo), is an indication that all the vertebrates are descended from some primitive creature with a notochord.

From several different fields then—from comparative anatomy, from paleontology, from biochemistry, from histology, cytology, and embryology—all signs at first whispered, then, as mid-nineteenth century approached, shouted that some sort of evolutionary view was a necessity. Some satisfactory mechanism for evolution simply had to be presented.

CHAPTER 6

Evolution

Natural Selection

The man who was to conceive an adequate evolutionary mechanism and establish it firmly in the minds of biologists was an English naturalist, Charles Robert Darwin (1809–82), grandson of the Erasmus Darwin mentioned earlier in the book (see page 39).

As a youth, Darwin tried to study medicine and later considered entering the Church, but neither career suited him. Natural history was a hobby of his and through his college days he became seriously interested in it as a career. In 1831, when the H.M.S. *Beagle* was about to set out for a voyage of scientific exploration around the globe, Darwin was offered the post of ship's naturalist, and accepted.

The voyage took five years and although Darwin suffered agonies of seasickness, it was the making of him as a naturalist of genius. Through him, moreover, the voyage of the *Beagle* became the most important exploring expedition in the history of biology.

Darwin had read Lyell's first volume on geology (see page 45) before starting out and had a clear realization of the antiquity of the earth and of the long ages through which life had had time to develop. Now, during the course of the voyage, he could not help but notice how species replaced each other—each succeeding species being slightly different from the one it replaced—as he traveled down the coast of South America.

Most striking of all were his observations, during a five-week stay, of the animal life of the Galapagos Islands, about 650 miles off the coast of Ecuador. In particular, Darwin studied a group of birds, called to this day "Darwin's finches." These finches, closely similar in many ways, were divided into at least fourteen species, not one of which existed on the nearby mainland or, as far as was known, anywhere else in the world. It seemed unreasonable to suppose that fourteen different species were created for this small and inconspicuous group of islands and for them alone.

Darwin felt, instead, that the mainland species of finch must have colonized the island long before and that, gradually over the eons, the descendants of those first finches evolved into different species. Some developed the habit of eating seeds of one sort, some of another, still others came to eat insects. For each way of life, a particular species developed a particular beak, a particular size, a particular scheme of organization. The original finch did not do this on the mainland because of competition from many other birds. In the Galapagos, however, the original finches found a relatively empty land, and there was room for the development of many varieties.

But one point, one key point, remained unanswered. What caused such evolutionary changes? What made one species of finch that ate seeds gradually become another that ate insects? Darwin could not accept an explanation of the Lamarckian type (see page 40); that is, a supposition that finches might have *tried* to eat insects and passed on the taste for it and an increased ability to manage it, to their offspring. Unfortunately, he had no other answer to substitute.

Then, in 1838, two years after his return to England, he came across a book entitled *An Essay on the Principle of Population* that had been written forty years earlier by an English economist, Thomas Robert Malthus (1766–1834). In this book, Malthus maintained that human population always increased faster than the food supply did and that eventually population had to be cut down by either starvation, disease, or war.

Darwin thought at once that this must hold for all other forms of life as well and that those of the excess population that were first cut down would be just those who were at a disadvantage in the competition for food. For instance, those first finches on the Galapagos must have multiplied unchecked to begin with and would surely have outstripped the supply of seeds they lived on. Some would have had to starve, the weaker ones first, or those less adept at finding seeds. But what if some just happened to be able to eat bigger seeds or get by on tougher seeds or found themselves able to swallow an occasional insect? Those which were not possessed of these unusual abilities would be held in check by starvation while those who could, however inefficiently, would find a new and untapped food supply and could then multiply rapidly until, in turn, their food supply began to dwindle.

In other words, the blind pressure of the environment would put a premium on differences, and would pile difference upon difference until separate species formed, each distinct from the other and from the common ancestor. Nature itself, so to speak, would select the survivors as the food supply grew short and by such "natural selection," life would branch out into infinite variety.

Furthermore, Darwin could see how the necessary changes took place. He bred pigeons to study the effects of artificial selection and therefore had personal experience with the breeding of odd varieties of domesticated animals. He could see that in any group of young there were random variations from one to another; variations in size, coloring, and abilities. It was through taking advantage of such variations, through deliberately breeding one and suppressing others, that over the generations man had developed improved breeds of cattle, horses, sheep, and poultry, and had bent dogs and goldfish into odd and amusing shapes to suit his fancy.

Could not Nature substitute for man and make the same sort of selection for its own purposes, much more slowly and over a much longer period, to be sure, but fitting animals to their environment rather than to man's tastes and demands?

Darwin also studied "sexual selection," in which the female of the species accepted the most flamboyant male, so that the almost ridiculously excessive peacock would develop. Then, too, he collected data on vestiges of organs that bespoke full-scale useful organs ages before. (As a dramatic example, consider that whales and snakes have scraps of bones that might once have formed parts of hip girdles and hind legs, a fact that almost forces us to believe that they must be descendants of creatures that once walked on legs.)

Darwin was a painstaking perfectionist and persisted in collecting and classifying his information endlessly. Finally, in 1844, he started writing on the subject. However, he did not get around to a thoroughgoing and definitive description of his theories for a decade thereafter. He finally launched into the main effort in 1856.

Meanwhile, in the Far East, another English naturalist, Alfred Russel Wallace (1823–1913), was considering the problem, too. Like Darwin, he had spent much of his life in travel, including a trip to South America between 1848 and 1852. In 1854, he sailed to the Malay peninsula and the East Indian islands. There he was struck by the sharp difference between the mammalian species of Asia and Australia. In later life, writing on this subject, he drew a line separating the lands in which these separate sets of species flourished. The line (still called "Wallace's Line") follows a deep-water channel that separates the large islands of Borneo and Celebes, and also separates the smaller islands of Bali and Lombok to the south. Out of this grew the notion of dividing the animal species into large continental and supercontinental blocs.

It seemed to Wallace that the mammals of Australia were more primitive and less efficient than those of Asia and that in any competition between the two the Australian mammals would perish. The reason that the Australian mammals survived at all was that Australia and the nearby islands had split off from the Asian mainland before the more advanced Asian species had developed. Such thoughts led him to speculate on evolution by natural selection. Exactly as in the case of Darwin, these speculations were brought to a head when he happened to read Malthus' book. Wallace was in the East Indies at the time, suffering from ague; employing his enforced leisure, he wrote out his theory in two days and sent the manuscript to Darwin for an opinion. (He did not know Darwin was working on the same project.) When Darwin received the manuscript he was thunderstruck at the close duplication of views. Lyell and others arranged to have some of Darwin's writings presented together with Wallace's paper and they were published in the "Journal of Proceedings of the Linnaean Society" in 1858.

The next year Darwin finally published his book *On the Origin of Species by Means of Natural Selection, or the Preservation of Favoured Races in the Struggle for Life*. It is usually known simply as *The Origin of Species*.

The learned world was waiting for the book. Only 1250 copies were printed and every copy was snapped up on the first day of publication. It went through printing after printing, and it is still being reprinted nowadays, a century later.

The Struggle over Evolution

Unquestionably, *The Origin of Species* was the most important book in the history of biology. A great many branches of the science suddenly made better sense when viewed from the standpoint of evolution by natural selection. The concept rationalized the gathering data on taxonomy, embryology, comparative anatomy, and paleontology. With Darwin's book, biology became more than a collection of facts; it became an organized science based upon a broad and extraordinarily useful theory.

But Darwin's book was hard for many to take. It upset some of the revered notions of men; in particular, it seemed to fly against the literal word of the Bible and to imply that God did not create the world and mankind. Even among those whose views were not particularly religious there were many who were repelled by a view that made the beautiful realm of life and even the miracle of man himself the product of the workings of a blind and unfeeling chance.

In England, the zoologist, Richard Owen (1804–92), the leader of the opposition, was a member of the latter group. He was a disciple of Cuvier and, like Cuvier, an expert in the reconstruction of extinct animals from fossil remnants. It was not the concept of evolution itself that he objected to, but the thought that it was brought about by chance. He preferred the notion of some inner urge.

Darwin himself did not actively fight for his own theory, for he was too gentle (and usually fancied himself too ill) to be much of a controversialist. However, the English biologist, Thomas Henry Huxley (1825–95), took up the cudgels on Darwin's behalf. Huxley was a terror on the lecture platform, and a gifted writer on science for the general public. He called himself "Darwin's bulldog" and he, more than anyone else, put evolution across for the common man.

Darwinism made little headway at first in France, where biologists remained under the antievolutionary spell of Cuvier for some decades. Germany, however, was much more fertile ground. The German naturalist, Ernst Heinrich Haeckel (1834–1919), went all the way, and a bit beyond, for Darwin. He saw the developing embryo as a virtual condensed motion picture of evolution. The mammal, for instance, began as a single cell, like a protozoon; developed into a two-germ-layered creature something like a jelly fish; then into a three-germ-layered creature something like a primitive worm. In the course of further development, the mammalian embryo produced and then lost the notochord of the primitive chordates; then produced and lost structures that seemed the beginning of fishlike gills. In this view, Haeckel was strenuously opposed by the older embryologist, Baer (see page 58), who had himself come to the edge of this view but would not accept Darwinism. Indeed, Haeckel's views proved too extreme, and modern biologists do not accept embryonic development as an entirely literal and faithful picture of the course of evolution.

In the United States, the American botanist, Asa Gray (1810–88), was the most active spokesman on behalf of Darwinism. He himself was a prominent religious layman which gave his point of view added force, since he could not be dismissed as an atheist. Opposed to him was the

Swiss-American naturalist, Jean Louis Rodolphe Agassiz (1807–73). Agassiz had made his scientific reputation on an exhaustive study of fossil fish, but with the general public, his more spectacular deed was that of popularizing the notion of the "Ice Ages." He was at home with the Alpine glaciers of his native Switzerland and was able to show that those glaciers moved slowly and that, in so doing, the embedded pebbles and detritus on their lower surface scraped and gouged the rocks over which they passed.

Agassiz found such grooved rocks, unmistakably glacier-gouged, in regions where no glaciers had ever existed in the memory of man. In the 1840s, he came to the conclusion that many thousands of years before, glaciers must have been widespread. In 1846, he came to the United States, primarily to lecture at first, but his interest in the natural history of the North American continent led him to decide to stay permanently. Here, too, he found signs of extensive ancient glaciation.

The Ice Age (now known to have existed as four separate Ice Ages within the last half-million years or so) was good evidence to the effect that the extreme uniformitarianism of Hutton and Lyell was not justified. There were catastrophes, after all. To be sure, these were not as sudden, as shattering, and as fatal as Cuvier's theories demanded, but they existed. Between his own Cuvierlike feelings and his natural piety, Agassiz found that he was incapable of accepting the Darwinian theory.

The Descent of Man

Naturally, the touchiest point about the Darwinian theory lay in its application to man himself. Darwin had skirted that point in The Origin of Species and Wallace, codiscoverer of the theory of natural selection, eventually came to maintain strongly that man himself was not subject to evolutionary forces. (He became a spiritualist in later life.) However, it was unreasonable to suppose that evolution would involve all species but *Homo sapiens* and there was slowly gathering evidence to the effect that man was indeed involved.

In 1838, for instance, a French archaeologist, Jacques Boucher de Crèvecoeur de Perthes (1788–1868), had dug up crude axes in northern France which, from their position in the strata, he could only judge as being many thousands of years old. Furthermore, they were clearly artificial and could have been made only by man. For the first time, there was undoubted evidence that not only the earth, but man himself, was far more ancient than the six thousand years which the Bible seemed to demand.

Boucher de Perthes published his findings in 1846 and the book created a furor. French biologists, still under the influence of the dead Cuvier, were hostile and refused to accept the implications of the find, even though archaeologists began to turn up more ancient tools in the 1850s. Finally, in 1859, a number of British scientists came to France, visited the spots where Boucher de Perthes had found his axes, and declared themselves on his side.

Four years later, Lyell, the geologist (see page 45), using Boucher de Perthe's findings as his evidence, published *The Antiquity of Man*, a book in which he not only strongly supported Darwinian notions but applied them specifically to man. Huxley (see page 66) also wrote a book taking up this position.

In 1871, Darwin openly joined the side in favor of human evolution with a second great book, *The Descent of Man*. Here he discussed man's vestigial organs as representing signs of evolutionary change. (There are a number of vestiges in the human body. The appendix is a remnant of an organ once useful for the storage of food which was thus allowed to undergo bacteria-induced breakdowns. There are four bones at the base of the spine that were once part of a tail; there are useless muscles designed for moving the ear, inherited from ancestors with ears that were movable; and so on.)

Nor was the evidence exclusively indirect. Ancient man himself appeared on the scene. In 1856, an old skull had been unearthed in the Neanderthal valley of Germany's Rhineland. It was clearly a human skull, but it was more primitive and apelike than any ordinary human skull would be. From the stratum in which it was located it had to be many thousands of years old. A controversy at once arose. Was it an early primitive form of man that later evolved into modern man, or was it simply an ordinary savage of ancient days, with a bone disease or a congenital skull malformation?

The German physician, Rudolf Virchow (1821–1902), maintained the latter, and he was an eminent authority. On the other hand, the French surgeon, Paul Broca (1824–80), the world's most renowned expert on skull structure at that time, insisted that no modern man, diseased or healthy, could possibly have a skull like that of the "Neanderthal man" and that the Neanderthal man was therefore an early form of man, quite different in some ways from modern man.

To settle matters required another find: some fossil discovery that would be truly intermediate between man and ape, a "missing link." Missing links were not unknown among the fossils. In 1861, for instance, the British Museum acquired a fossil of a creature that was clearly a bird, for there were imprints of feathers in the rock, yet it also possessed a lizardlike tail and lizardlike teeth. It was taken at once as the best possible evidence that birds had descended from reptiles.

The search for a specifically human missing link, however, was fruitless for some decades. Success came at last to a Dutch paleontologist, Marie Eugène François Thomas Dubois (1858–1940). Dubois was on fire with the hope of finding the missing link. To him, it seemed that primitive manlike creatures must be searched for in those areas of the world where apes still abounded; that is, either in Africa, home of the gorilla and chimpanzee; or in southeast Asia, home of the orangutan and gibbon.

In 1889, he was commissioned by the Dutch Government to search for fossils in Java (then a Dutch possession) and he threw himself into the task with great fervor. Within a matter of a very few years, he discovered a skullcap, a thighbone, and two teeth of what was undoubtedly a primitive man. The skullcap was considerably larger than that of any living ape, and yet considerably smaller than that of any living man. The teeth, too, were intermediate between ape and man. Dubois called the creature to which these bony remnants had belonged, *"Pithecanthropus erectus"* (the erect ape-man), and published the details in 1894.

Again, there was great controversy, but other similar finds have been made in China and Africa, so that a number of "missing links" are now known to have existed. No reasonable doubt remains of the fact of human evolution or of evolution in general. Much antievolution sentiment existed into the twentieth century and some, indeed, exists down to the present day, but this is largely among the fundamentalist religious sects who insist on the literal words of the Bible. It is difficult to imagine a reputable biologist as being antievolutionist in sentiment now.

Offshoots of Evolution

If the antievolutionists were in error, there was error also in overenthusiastic acceptance of evolution in areas where the theory did not apply. Thus, an English philosopher, Herbert Spencer (1820–1903), who had had evolutionary ideas even before Darwin's book was published, seized upon that book gladly. He added it to his own speculations on human society and culture and in this way became a pioneer in the study of *sociology*.

Spencer felt that human society and culture had begun

at some homogeneous and simple level and had evolved to its present heterogeneous and complex state. He popularized the term "evolution" (which Darwin hardly used) and the phrase "survival of the fittest." It seemed to Spencer that human individuals were in continual competition among themselves, with the weaker necessarily going to the wall. Spencer considered this to be an inevitable accompaniment of evolutionary advance and argued, in 1884, that people who were unemployable or who were otherwise burdens on society should be allowed to die rather than made objects of help and charity. Kindness and soft-heartedness, he maintained, impeded evolutionary advance and was harmful in the long run.

This, however, was using the language of evolution inappropriately, for the Darwinian mechanism of natural selection required long ages. As a matter of fact, the only way in which Spencer could justify the rapid changes in man's history was to adopt a form of inheritance of acquired characteristics after the fashion of Lamarck (see page 40). Spencer also had to ignore the fact that there might be survival value in a society that took care of its aged and infirm, since the individual members might then be more devoted to it. In fact, the history of civilization records the long-range triumph of social co-operation in agriculture and industry over the dog-eat-dog individualism of the huntsman and nomad.

Nevertheless, Spencerian evolutionary thought had its effect on history, for during the decades prior to World War I, it gave extreme nationalists and militarists a chance to speak of war being "good," since it insured the survival of those most fit. Fortunately, such romantic illusions about the despicable business of war no longer exist.

Another controversial turning was taken by the English anthropologist, Francis Galton (1822–1911), who was a first cousin of Darwin's. Galton spent his early years as an explorer and meteorologist, but after the appearance of his cousin's book, he turned to biology. He was particularly interested in the study of heredity and was the first to stress the importance of studying identical twins, where hereditary influences might be considered equal so that differences could be attributed to environment only.

By studying the occurrence of high mental ability in families, Galton was able to present evidence in favor of the view that mental ability was inherited. He felt, therefore, that human intelligence and other desirable characteristics, too, could be accentuated by proper breeding, while undesirable characteristics were bred out of the race. In 1883, he gave the name "eugenics" (from Greek words meaning "good birth") to the study of methods whereby this could best be brought about. In his will, he left a bequest to be used for establishing a laboratory devoted to research in eugenics.

Unfortunately, as more and more information has been gathered concerning the mechanism of heredity, biologists have become less and less confident that the improvement of the race by selective breeding (purposeful evolution, so to speak) is a simple matter. In fact, it seems certain that it is a very complicated matter. While eugenics remains a legitimate concern of biology, the loudest so-called eugenicists are among small groups of nonscientists who use the language of science to beat their private tom-toms of racism.

CHAPTER 7

The Beginnings of Genetics

The Gap in Darwinian Theory

The reason why it was so easy to misapply evolutionary theory was that the nature of the hereditary mechanism was not understood in the nineteenth century. Spencer could imagine rapid changes in human behavior, and Galton could imagine improving the race by a quick and easy program of selective breeding, out of an ignorance they shared with biologists generally.

In fact, the lack of understanding of the nature of the hereditary mechanism was the most deplorable weakness of Darwinian theory. Put briefly, the weakness was this: Darwin supposed that there were continual random variations among the young of any species and that some variations would better fit an animal for its environment than would others. The young giraffe who happened to grow the longest neck would be the best fed.

But how could one be certain that the longest neck would be passed on? The giraffe was not likely to seek out a long-necked mate; it was as likely to find a short-necked one. All Darwin's experiences with the breeding of animals led him to suppose that there was a blending of characteristics when extremes were crossed so that a long-necked giraffe mated with a short-necked giraffe would give rise to young with medium-length necks.

In other words, all the useful, well-fitting characteristics that were introduced by random variation would average out into an undistinguished middle ground as a result of equally random mating and there would be nothing upon which natural selection could seize to bring about evolutionary changes.

Some biologists made stabs at explaining this away, but without much success. The Swiss botanist, Karl Wilhelm von Nägeli (1817–91), was an enthusiastic supporter of Darwinism and recognized the difficulty. He supposed, therefore, that there must be some inner push that drove evolutionary changes in a particular direction.

Thus, the horse, as was known from the fossil record, was descended from a dog-sized creature with four hoofs on each foot. Through the ages the descendants grew continually larger and lost one hoof after another until the modern large, one-hoofed horse was developed. Nägeli felt that there was an inner drive that moved the developing horse constantly in the direction of larger size and fewer toes and that this might be continued even to the point of harm so that horses might become too large and clumsy for their own good. Unable to escape from their enemies, they would then decline progressively in numbers and become extinct.

This theory is called "orthogenesis" and it is *not* accepted by modern biologists. However, its existence in Nägeli's mind proved unexpectedly harmful as we shall now see.

Mendel's Peas

The solution to the problem, one which is now accepted, arose through the work of an Austrian monk and amateur botanist, Gregor Johann Mendel (1822–84). Mendel was interested in both mathematics and botany and, combining the two, studied peas statistically for eight years, beginning in 1857.

Carefully, he self-pollinated various plants, making sure in this way that if any characteristics were inherited, they would be inherited from only a single parent. As carefully, he saved the seeds produced by each self-pollinated pea plant, planted them separately, and studied the new generation.

He found that if he planted seeds from dwarf pea plants, only dwarf pea plants sprouted. The seed produced by this second generation also produced only dwarf pea plants. The dwarf pea plants "bred true."

Seeds from tall pea plants did not always behave in quite this way. Some tall pea plants (about a third of those in his garden) did indeed breed true, producing tall pea plants generation after generation. The rest, however, did not. Some seeds from these other tall plants produced tall plants and others produced dwarf plants. There were always about twice as many tall plants produced by these seeds as dwarf plants. Apparently, then, there were two kinds of tall pea plants, the true breeders and the nontrue breeders.

Mendel then went a step further. He crossbred dwarf plants with true-breeding tall plants and found that every resulting hybrid seed produced a tall plant. The characteristic of dwarfness seemed to have disappeared.

Next, Mendel self-pollinated each hybrid plant and studied the seeds produced. All the hybrid plants proved to be nontrue breeders. About one quarter of their seeds grew into dwarf plants, one quarter into true-breeding tall plants, and the remaining half into nontrue-breeding tall plants.

Mendel explained all this by supposing that each pea plant contained two factors for a particular characteristic such as height. The male portion of the plant contained one and the female portion contained the second. In pollination, the two factors combined and the new generation had a pair (one from each parent if they had been produced by a cross between two plants). Dwarf plants had only "dwarf" factors, and combining these by either cross-pollination or self-pollination, produced only dwarf plants. True-breeding tall plants had only "tall" factors and combinations produced only tall plants.

If a true-breeding tall plant were crossed with a dwarf plant, "tall" factors would be combined with "dwarf" factors, and the next generation would be hybrids. They would all be tall, because tallness was "dominant," drowning out the effect of the "dwarf" factor. The "dwarf" fac-

T – tall
d – dwarf
Td – non-true-breeding or hybrid tall

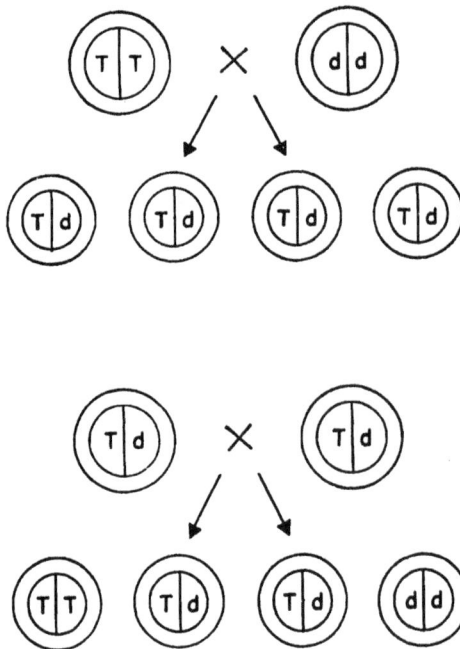

FIGURE 3. Diagrammatic explanation of Mendel's work with tall and dwarf pea plants. The top illustration is the crossing of a true tall plant with a dwarf plant, resulting in hybrid (or non-true-breeding) tall plants. Below, the crossing of hybrid tall plants which results in true tall plants, dwarf plants, and hybrid tall plants, in proportions of 1:1:2.

tor would, however, still be there. It would not have vanished.

If such hybrids are either cross-pollinated or self-pollinated, they prove to be nontrue breeding because they possess both factors which can be combined in a variety of ways (dictated by chance alone). A "tall" factor might combine with another "tall" factor to produce a true-breeding tall plant. This would happen one quarter of the time. A "dwarf" factor might combine with another "dwarf" factor to produce a dwarf plant. This would also happen a quarter of the time. The remaining half of the time, a "tall" factor would combine with a "dwarf" factor, or a "dwarf" factor with a "tall" factor, to produce non-true-breeding tall plants.

Mendel went on to show that a similar explanation would account for the manner of inheritance of characteristics other than height. In the case of each set of characteristics he studied, crossing two extremes did *not* result in a blend into intermediateness. Each extreme retained its identity. If one disappeared in one generation, it showed up in the next.

This was of key importance to the theory of evolution (although Mendel never thought of applying his ideas to that theory), for it meant that random variations produced in species in the course of time did not average out after all but kept appearing and reappearing until natural selection had made full use of them.

The reason why characteristics often seemed to become intermediate after random mating is that most "characteristics" casually observed by breeders of plants and animals are really combinations of characteristics. The different components can be inherited independently and while each is inherited in a yes-or-no manner, the over-all result of some yeses and some noes is to lend an appearance of intermediacy.

Mendel's findings also affected the notions of eugenics. It was not as easy to eradicate an undesirable characteristic as one might think. It might not appear in one generation, and yet would crop up in the next. Selective breeding would have to be more subtle and more prolonged than Galton imagined.

However, the world was not to know of all this just yet. Mendel wrote up the results of his experiments carefully, but, conscious of his own status as an unknown amateur, felt it would be wise to obtain the interest and sponsorship of a well-known botanist. In the early 1860s, therefore, he sent his paper to Nägeli. Nägeli read the paper and commented upon it coldly. He was not impressed by theories based on counting pea plants. He preferred obscure and wordy mysticism, such as his own orthogenesis.

Mendel was disheartened. He published his paper in 1866, but did not continue his research. Moreover, without Nägeli's sponsorship, the paper lay disregarded and unnoticed. Mendel had founded what we now call *genetics* (the study of the mechanism of inheritance) but neither he nor anyone else knew it at the time.

Mutation

Another problem arose, in connection with evolution, during the latter half of the nineteenth century. The long time scale of Earth history was suddenly imagined to be much shorter as a result of new findings of physics. With the enunciation of the law of conservation of energy, the question had arisen as to where the sun's energy came from. Nothing was known, at the time, of radioactivity or of nuclear energy, so all nineteenth-century explanations were insufficient to account for the existence of the sun in its present state for more than, at most, a few tens of millions of years.

This was simply not enough time for evolution to proceed in a normal Darwinian fashion, and some biologists such as Nägeli and Kölliker wondered if evolution might not proceed by jumps. Though the short time scale proved wrong and though there turned out, in the end, to be no need at all to skimp on the time allotted for evolution, the suggestion of evolution by jumps proved fruitful.

A Dutch botanist, Hugo de Vries (1848–1935), who was one of those who speculated on evolution by jumps, came across a colony of American evening primroses growing in a waste meadow. These plants had been introduced into the Netherlands some time before and De Vries's botanical eye was caught by the fact that some of these primroses, though presumably descended from the same original plant as the rest, were widely different in appearance.

He brought them back to his garden, bred them separately, and gradually came to the same conclusions that Mendel had reached a generation earlier. He found that individual characteristics were passed along from generation to generation without blending and becoming intermediate. What's more, every once in a while, a new variety of plant, differing markedly from the others, would appear, and this new variety would perpetuate itself in future generations. De Vries called such a sudden change a "mutation" (from the Latin word for "change") and recognized the fact that here before his eyes was evolution by jumps. (Actually, the sort of mutation exhibited by the evening primrose was a rather simple sort not involving actual changes in the heredity factors themselves. Soon, however, true mutations, involving such changes, came to be studied.)

This sort of thing had always been known to herdsmen and farmers, who had frequently seen the production of "freaks" or "sports." Some sports had even been put to use. For instance, a short-legged sheep (a mutation) had appeared in New England in 1791. Since it was too short-legged to jump over even low fences, it was useful, and was bred and preserved. However, herdsmen do not usually draw theoretical conclusions from their observations, nor do scientists usually involve themselves with the mechanics of herding.

With De Vries, however, the phenomenon and the scientist finally met. By 1900, when he was ready to publish his findings, a check through previous work on the subject revealed Mendel's thirty-four-year-old papers to his astonished eyes.

Unknown to De Vries and to each other, two other botanists, the German, Karl Erich Correns (1864–1933) and the Austrian, Erich Tschermak von Seysenegg (1871–   ), had that same year reached conclusions very similar to those of De Vries. Each then looked through previous writings on the subject and found Mendel's papers.

All three, De Vries, Correns, and Tschermak von Seysenegg published their works in 1900 and each of the three cited Mendel's work and listed their own work simply as confirmation. So it is that we speak of the Mendelian laws of inheritance. The combination of these laws with De Vries's discovery of mutations described the manner in which variations originated and were preserved. The shortcomings in Darwin's original theory were thus removed.

Chromosomes

The Mendelian laws were more significant in 1900 than they were in 1866 because in the interim important new discoveries had been made concerning cells.

Those who observed cells during the eighteenth and early nineteenth centuries did not see much, even with improved microscopes. The cell was a virtually transparent body and so was the material within it. Consequently, it seemed a more-or-less featureless blob, and biologists had to be content to describe its over-all size and shape, and no more. Some occasionally made out a denser region (now called the "cell nucleus") near its center, but the first to recognize this as a regular feature of cells was the

Scottish botanist, Robert Brown (1773–1858), who made this suggestion in 1831.

Seven years later, when Schleiden advanced the cell theory (see page 57), he attributed considerable importance to the cell nucleus. He felt that it was connected with cell reproduction and that new cells budded out of the nuclear surface. By 1846, Nägeli was able to show that this was wrong. However, Schleiden's intuition did not lead him altogether astray; the nucleus *was* involved in cell reproduction. Knowledge concerning the details of this involvement, however, had to await some new technique for viewing the cell's interior.

The technique came by way of organic chemistry. Following the lead of Berthelot, organic chemists were rapidly learning how to prepare organic chemicals that did not exist in nature. Many of these were brightly colored and, indeed, the 1850s saw the beginnings of the gigantic "synthetic dye" industry.

Now if the interior of the cell were heterogeneous, then it was quite possible that some parts might react with a particular chemical and absorb it, while other parts might not. If the chemical were a dye, the result would be that some parts of the cell would become colored, while others would remain colorless. Detail unseen before would spring into view, thanks to such "stains."

A number of biologists experimented in this fashion and one of those who was outstandingly successful was the German cytologist, Walther Flemming (1843–1905). Flemming studied animal cells and found that scattered within the cell nucleus were spots of material that strongly absorbed the dye he was working with. They stood out brightly against the colorless background. Flemming called this absorptive material "chromatin" (from the Greek word for "color").

When Flemming dyed a section of growing tissue, he killed the cells, of course, but each was caught at some stage of division. In the 1870s Flemming was able to work

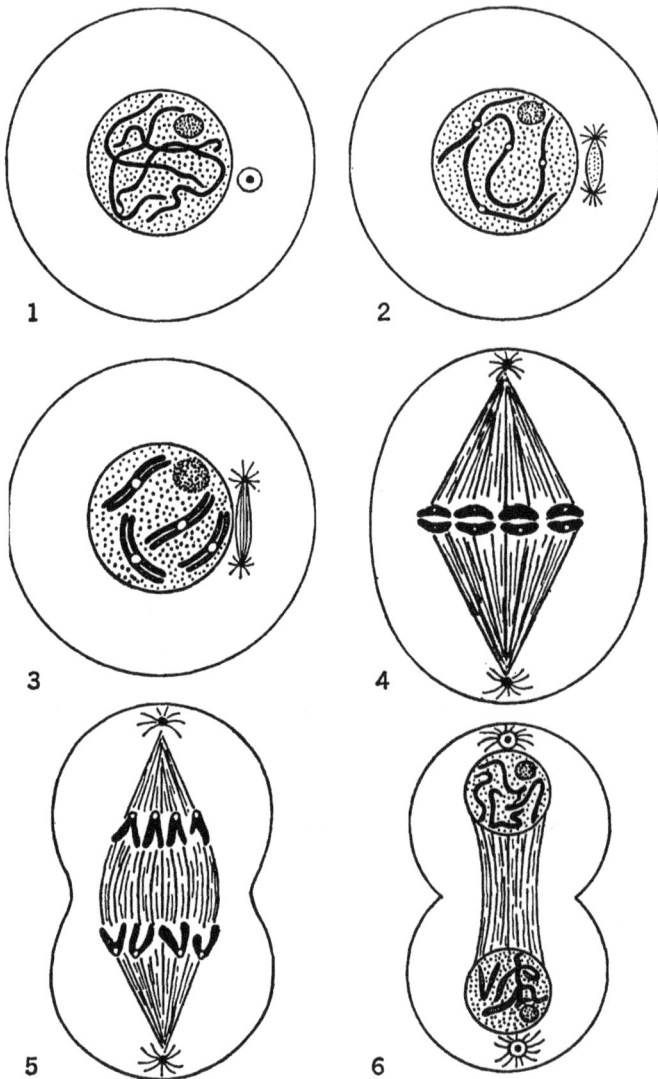

FIGURE 4. The stages of mitosis. (1) Chromosomes form in the nucleus in the first stage of mitosis. (2) Chromosomes begin to split into two identical halves and the aster (the small white circle outside the nucleus) spreads to opposite sides of the cell. (3) Chromosomes have doubled but remain joined at the

out the changes in the chromatin material that accompanied the progressive changes of cell division.

He found that as the process of cell division began, the chromatin material coalesced into short threadlike objects which later came to be called "chromosomes" ("colored bodies"). Because these threadlike chromosomes were so characteristic a feature of cell division, Flemming named the process "mitosis" (from a Greek word for "thread").

Another change that accompanied the beginning of mitosis involved the "aster" (a Greek word meaning "star"). This received the name because it was a tiny dotlike object surrounded by fine threads radiating from it like the conventional rays drawn from a star. The aster divided, the two parts separating and moving to opposite sides of the cell. The fine rays passing from one aster to the other seemed to entangle the chromosomes which were grouping along the midplane of the cell.

At the crucial moment of cell division, each chromosome produced a replica of itself. The double chromosomes then pulled apart, one chromosome of each doublet going to one end of the cell and the second chromosome to the other. The cell then divided, a new membrane forming down the middle. Where there had previously been one cell, there were now two daughter cells, each with an amount of chromatin material equal (thanks to the doubling of the chromosomes) to that which had originally been present in the mother cell. Flemming published these findings in 1882.

The work was carried further by the Belgian cytologist, Eduard van Beneden (1846–1910). In 1887, he was able to demonstrate two important points about chromosomes. First, he presented evidence to show that their

center. (4) Chromosomes are lined up and asters have moved to opposite poles. (5) Asters pull chromosomes apart. (6) Cell begins to lengthen and ultimately will form, two new identical cells each with its own nucleus and an identical amount of chromatin as was in the mother cell in the first stage.

number was constant in the various cells of an organism, and that each species seemed to have a characteristic number. (It is now known, for instance, that each human cell contains forty-six chromosomes.)

Further, Van Beneden discovered that in the formation of the sex cells, the ova (egg cells), and spermatozoa, the division of chromosomes during one of the cell divisions was *not* preceded by replication. Each egg and sperm cell, therefore, received only half the usual count of chromosomes.

Once Mendel's work had been discovered by De Vries, all this work on chromosomes was suddenly illuminated. The American cytologist, Walter S. Sutton (1876–1916), pointed out in 1902 that the chromosomes behaved liked Mendel's inheritance factors. Each cell has a fixed number of pairs of chromosomes. These carry the capacity to produce physical characteristics from cell to cell, for in each cell division, the number of chromosomes is carefully conserved; each chromosome producing a replica of itself for the use of the new cell.

When an egg cell (or a sperm cell) is formed, each receives only half the usual chromosome number (one of each pair). When the fertilized ovum is formed from the union of sperm and ovum, the correct total number of chromosomes is restored. As the fertilized ovum divides and redivides to form an independently living organism, the number of chromosomes is again carefully conserved. In the new organism, however, one of each pair of chromosomes comes from the mother via the egg cell, while the second of each pair comes from the father via the sperm cell. This shuffling of chromosomes with each generation tends to bring to light those recessive characteristics earlier drowned out by a dominant characteristic. The ever-new combinations further produce over-all variations of characteristics upon which natural selection can seize.

As the twentieth century dawned, then, a sort of climax

had been reached in evolution and genetics. This, however, was only to serve as a prelude to new and even more startling advances.

CHAPTER 8

The Fall of Vitalism

Nitrogen and the Diet

However unsettling Darwin's theory of evolution by natural selection might have been to many of mankind's settled beliefs, it did, viewed properly, enhance the marvel of life. From very simple beginnings, life had striven endlessly, under the stress of environment, to achieve ever greater complexity and efficiency. There was nothing to compare with that in the changeless world of the inanimate. Mountains might rise but there had been other mountains eons before; life forms, on the other hand, were ever new, ever different.

Darwinian theory might therefore be interpreted at first blush as favorable to vitalism, to the great barrier thrown up in men's minds between life and nonlife. And indeed vitalism reached a new height of popularity in the latter half of the nineteenth century.

The great danger to nineteenth-century vitalism lay in the advances made by organic chemists (see page 53). Against this, however, the vitalists raised the protein molecule as a shield and down almost to the very end of the century, that shield held firm. The biochemistry of the nineteenth century was very largely concerned with that protein molecule.

The importance of protein to life was first made completely clear by the French physiologist, François Magendie (1783–1855). The economic dislocations brought on by the Napoleonic wars had brought a period of food scarcity, and the condition of the poor was worse than usual. Governments were beginning to feel a responsibility for the condition of the people, and a commission was appointed, with Magendie at its head, to investigate whether a nourishing food could be made out of something as cheap and available as gelatin.

Magendie began, in 1816, by feeding dogs on a protein-free diet, one that contained only sugar, olive oil, and water. The animals starved to death. Calories alone were not sufficient; protein was a necessary component of the diet. Furthermore, not all proteins were equally useful. Unfortunately, where gelatin was the only protein in the diet, the dogs still died. Thus was founded the modern science of *nutrition*, the study of diet and its connection with life and health.

Proteins differed from the carbohydrates and lipids in that the former contained nitrogen and the latter did not. For that reason, interest focused on nitrogen as a necessary component of living organisms. The French chemist, Jean Baptiste Boussingault (1802–87), set out in the 1840s to study the nitrogen requirements of plants. He found that some plants, such as the legumes (peas, beans, etc.), could grow readily in nitrogen-free soil while being watered with nitrogen-free water. Not only did they grow, but their nitrogen content increased steadily. The only conclusion he could come to was that these plants gained their nitrogen from the air. (We now know that it is not the plants themselves that do this, but certain strains of "nitrogen-fixing bacteria" growing in root nodules that do so.)

Boussingault, however, went on to show that animals could obtain no nitrogen from the air, but only from food. He sharpened Magendie's rather qualitative studies by actually analyzing the nitrogen content of some foods and comparing the rate of growth with the nitrogen content. There was a direct relationship, provided a single food was used as nitrogen source. However, some foods were more efficient than others at bringing about growth with a given nitrogen content. The conclusion could only be that some proteins were more useful, nutritionally, to the body than others were. The reason for this remained obscure till the end of the century, but, by 1844, Boussingault could, on purely empirical grounds, list the relative usefulness of various foods as sources of protein.

This was carried further by the German chemist, Justus von Liebig (1805–73), who over the following decade prepared detailed lists of this sort. Liebig leaned strongly toward mechanism, and he applied this viewpoint to the problems of agriculture. He believed that the reason for loss of soil fertility after years of farming lay in the gradual consumption of certain minerals in the soil which were necessary for plant growth. Plant tissues contained small quantities of sodium, potassium, calcium, and phosphorus, and these had to come from soluble compounds in the soil, which the plant could absorb. It had been customary from time immemorial to bolster soil fertility by the addition of animal refuse, but to Liebig this did not signify the addition of something "vital" to the soil, but merely that of the mineral content of the wastes to replenish that which had been taken out of the soil. Why not add the minerals themselves, pure, clean, and odorless, and do away with the necessity of dealing with wastes?

He was the first to experiment with chemical fertilizers. At first, his products were failures because he relied too heavily on Boussingault's finding that some plants obtain nitrogen from the air. When Liebig realized that most plants, after all, obtain nitrogen from soluble nitrogen compounds ("nitrates") in the soil, he added these to his mixture and produced useful fertilizers. Between them, Boussingault and Liebig founded *agricultural chemistry*.

Calorimetry

Liebig, as a good mechanist, believed that carbohydrates and lipids were the fuels of the body just as they would be fuel for a bonfire if thrown into one. This marked an advance over Lavoisier's views of a half-century earlier (see page 47). Lavoisier had then been able to speak of carbon and hydrogen only, whereas now one could speak, more specifically, of the carbohydrates and lipids which were made up of carbon and hydrogen (plus oxygen).

Liebig's views naturally encouraged attempts to determine whether the amount of heat obtained from such fuel in the body was really the same as that obtained if the carbohydrates and fats were simply burned in ordinary fashion outside the body. Lavoisier's crude experiments had indicated the answer would be "yes," but techniques had been refined since his day and it was now necessary to put the question more rigorously.

Devices to measure the heat released by burning organic compounds were developed in the 1860s. Berthelot (see page 51) utilized such a device ("calorimeter") to measure the heat produced by hundreds of such reactions. In an ordinary calorimeter, such as that which Berthelot used, a combustible substance is mixed with oxygen in a closed chamber and the mixture is exploded by means of a heated electrified wire. The chamber is surrounded by a water bath. The water absorbs the heat produced in the combustion and from the rise in the temperature of the water, one can determine the amount of heat that has been released.

In order to measure the heat produced by organisms, a calorimeter must be built large enough to hold that organism. From the amount of oxygen the organism consumes and the amount of carbon dioxide it produces, the quantity of carbohydrate and lipid it "burns" can be calculated. The body heat produced can be measured, again by the rise in temperature of a surrounding water jacket. That heat can then be compared with the amount that would have been obtained by the ordinary burning of the same quantity of carbohydrate and lipid outside the body.

The German physiologist, Karl von Voit (1831–1908), a student of Liebig's, together with the German chemist, Max von Pettenkofer (1818–1901), designed calorimeters large enough to hold animals and even human beings. The measurements they made seemed to make it quite likely that living tissue had no ultimate energy source other than what was available in the inanimate universe.

Voit's pupil, Max Rubner (1854–1932), carried matters further and left no possibility of any remaining doubt. He measured the nitrogen content of urine and feces and carefully analyzed the food he fed his subjects in order that he might draw conclusions as to the proteins as well as the carbohydrates and lipids. By 1884, he was able to show that carbohydrates and lipids were not the only fuels of the body. Protein molecules could also serve as fuel after the nitrogen-containing portions were stripped away. Allowing for protein fuel, Rubner was able to make his measurements that much more accurate. By 1894, he was able to show that the energy produced from foodstuffs by the body was precisely the same in quantity as it would have been if those same foodstuffs had been consumed in a fire (once the energy content of urine and feces were allowed for). The law of conservation of energy held for the animate as well as the inanimate world, and in that respect at least there was no room for vitalism.

These new measurements were put to work on behalf of medicine. A German physiologist, Adolf Magnus-Levy (1865–1955), beginning in 1893, measured the minimum rate of energy production ("basal metabolic rate" or "BMR") in human beings and found significant changes in diseases involving the thyroid gland. Thereafter, measurements of BMR became an important diagnostic device.

Fermentation

The advances in calorimetry in the latter half of the nineteenth century left the core of vitalism untouched, however. Man and the rock he stood upon might both be composed of matter but an impassable line was drawn between forms of matter—first, organic versus inorganic and, when that failed, protein versus nonprotein.

In the same way, the total energy available might be the same for life and nonlife, but surely there was an impassable line between the methods whereby such energy was made available.

Thus, outside the body, combustion was accompanied by great heat and light. It proceeded with violence and rapidity. The combustion of foodstuffs within the body, however, produced no light and little heat. The body remained at a gentle 98.6° F. and combustion within it proceeded slowly and under perfect control. In fact, when the chemist tried to duplicate a reaction characteristic of living tissue he was generally forced to use drastic means: great heat, an electric current, strong chemicals. Living tissue required none of this.

Is this not a fundamental difference?

Liebig maintained it was not and pointed to fermentation as an example. From prehistoric times, mankind had fermented fruit juices to make wine and steeped grain to make beer. They had used "leaven" or yeast (as it is more often called) to make dough undergo changes that caused it to puff up with bubbles and make soft, tasty bread.

These changes involve organic substances. Sugar or starch is converted to alcohol and this resembles reactions that go on in living tissue. Yet fermentation does not involve strong chemicals or drastic means. It proceeds at room temperature and in a quiet, slow manner. Liebig maintained that fermentation was a purely chemical process that did not involve life. He insisted it was an example of a change that could take place life fashion, yet without life.

To be sure, since Van Leeuwenhoek's time (see page 29), yeast was known to consist of globules. The globules showed no obvious signs of life, but in 1836 and 1837, several biologists, including Schwann (see page 57), had caught them in the act of budding. New globules were being formed and this seemed to be a sure indication of life. Biologists began to speak of yeast cells. This, however, Liebig did not allow. He did not accept the living nature of yeast.

A French chemist, Louis Pasteur (1822–95), took up the cudgels against the redoubted Liebig. In 1856, he was called in for consultation by the leaders of France's wine industry. Wine and beer often went sour as they aged, and millions of francs were lost as a result. Was there not something a chemist could do?

Pasteur turned to the microscope. He found almost at once that when wine and beer aged properly, the liquid contained tiny spherical yeast cells. When wine and beer turned sour, however, the yeast cells present were elongated. Clearly, there were two types of yeast: one which produced alcohol and one which, more slowly, soured the wine. Heating the wine gently would kill the yeast cells and stop the process. If this were done at the right moment, after the alcohol had formed and before the souring had set in, all would be well. And all was!

In the process, Pasteur made two points quite plain. First, the yeast cells *were* alive, since gentle heat destroyed their ability to bring about fermentation. The cells were still there; they had not been destroyed, only the life within them had. Second, only living yeast cells, not dead ones, could bring about fermentation. The controversy between himself and Liebig ended in a clear victory for Pasteur and vitalism.

Pasteur went on to perform a famous experiment in connection with spontaneous generation, a subject on which the vitalist position had hardened since Spallanzani's time (see page 34). Biblical evidence in favor of spontaneous generation was now discounted and indeed religious leaders welcomed the disproof of spontaneous generation since that would reserve the formation of life to God alone. It was the mechanists of the mid-nineteenth century who, in some cases passionately, supported spontaneous generation.

Spallanzani had shown that if meat broth were sterilized and sealed away from contamination no life forms would develop in it. Those who were at the time in opposition maintained that heat had destroyed a "vital principle" in the air within the sealed chamber. Pasteur therefore devised an experiment in which ordinary unheated air would not be kept away from the meat broth.

In 1860, he boiled and sterilized meat broth and left it open to the ordinary atmosphere. The opening, however, was by way of a long, narrow neck, shaped like an S, lying on its side. Although unheated air could thus freely penetrate into the flask, any dust particles present would settle to the bottom of the S and did not enter the flask.

Under such conditions, the meat broth bred no organisms, but if the neck were removed, contamination followed shortly. It was not a question of heated or unheated air, of a "vital principle" destroyed or undestroyed. It was a matter of dust, some of which consisted of floating microorganisms. If these fell into the broth, they grew and multiplied; if not, not.

The German physician, Rudolf Virchow (see page 69), added to this as a result of his own observations. In the 1850s, he studied diseased tissue intensively (and is therefore considered the founder of the modern science of *pathology*, the study of diseased tissue) and demonstrated that the cell theory applies to it as well as to normal tissue.

The cells of diseased tissue, he showed, were descended from normal cells of ordinary tissue. There was no sudden break or discontinuity; no eruption of abnormal cells from nowhere. In 1855, Virchow epitomized his notion of the cell theory by a pithy Latin remark which can be translated as "All cells arise from cells."

He and Pasteur together had thus made it quite clear that every cell, whether it was an independent organism or part of a multicellular organism, implied a pre-existing cell. Never had life seemed so permanently and irretrievably walled off from nonlife. Never had vitalism seemed so strong.

Enzymes

Yet if life forms could perform chemical feats that could not be performed in inanimate nature, these had to be accomplished by some material means (unless one were willing to depend on the supernatural, which nineteenth-century scientists were not willing to do). The nature of the material means slowly came into view.

Even in the eighteenth century, chemists had observed that a reaction could sometimes be hastened by the introduction of a substance that did not, to all appearances, take part in the reaction. Observations of this sort accumulated and attracted serious attention in the early nineteenth century.

A Russian chemist, Gottlieb Sigismund Kirchhoff (1764–1833), showed in 1812 that if starch were boiled with dilute acid, the starch broke down to a simple sugar, glucose. This would not happen if the acid were absent and yet the acid did not seem to take part in the reaction, for none of it was used up in the breakdown process.

Four years later, the English chemist, Humphry Davy (1778–1829), found that platinum wires encouraged the combination, at ordinary temperatures, of various organic vapors, such as alcohol, with oxygen. The platinum certainly did not seem involved in the reaction.

These and other examples came to the attention of Berzelius (see page 50) who wrote on the subject in 1836 and who suggested the name "catalysis" for the phenomenon. This is from Greek words meaning "to break down" and possibly refers to the acid-catalyzed breakdown of starch.

Ordinarily, alcohol burns in oxygen only after being heated to a high temperature at which its vapors ignite. In the presence of the platinum catalyst, however, the same reaction takes place without preliminary heating. It could therefore be argued that the chemical processes in living tissue proceed, as they do, under very gentle conditions, because certain catalysts are present in tissue that are not present in the inanimate world.

Indeed, in 1833, shortly before Berzelius dealt with the subject, the French chemist, Anselme Payen (1795–1871), had extracted a substance from sprouting barley which could break down starch to sugar even more readily than acid could. He named it "diastase." Diastase and other similar substances were named "ferments" because the conversion of starch to sugar is one of the preliminaries in the fermentation of grain.

Ferments were soon obtained from animal organisms as well. The first of these was from digestive juice. Réaumur (see page 46) had shown that digestion was a chemical process, and in 1824, the English physician, William Prout (1785–1850), had isolated hydrochloric acid from stomach juices. Hydrochloric acid was a strictly inorganic substance and this was a surprise to chemists generally. However, in 1835, Schwann, one of the founders of the cell theory (see page 57), obtained an extract from stomach juice that was not hydrochloric acid but which decomposed meat even more efficiently than the acid did. This, which Schwann named "pepsin" (from a Greek word meaning "to digest") was the true ferment.

More and more ferments were discovered and it became quite apparent in the latter half of the nineteenth century that these were the catalysts peculiar to living tissue; the catalysts that made it possible for organisms to do what chemists could not. Proteins remained the vitalist shield for there were many reasons for believing that these ferments were protein in nature (though this was not definitely demonstrated until the twentieth century).

It was a strain on the vitalist position, however, that some ferments worked as well outside the cell as inside. The ferments isolated from digestive juices performed their digestive work very well in a test tube. One might suspect that if one could obtain samples of all the various ferments, then any reaction that went on in a living organism could be duplicated in the test tube and without the intervention of life, since the ferments themselves (at least, those studied) were indubitably nonliving. What's more, the ferments followed the same rules obeyed by inorganic catalysts, such as acids or platinum.

The vitalist position, then, was that ferments from digestive juices did their work outside the cells anyway. A digestive juice poured into the intestines might as well be poured into a test tube. The ferments that remained within the cell and did their work *only* within the cell were a different matter. Those, insisted the vitalists, were beyond the grip of the chemist.

Ferments came to be divided into two classes: "unorganized ferments" or those that worked outside cells, like pepsin; "organized ferments" or those that worked inside cells only, like those that enabled yeast to convert sugar into alcohol.

In 1876, the German physiologist, Wilhelm Kühne (1837–1900), suggested that the word, ferment, be reserved for those processes requiring life. Those ferments which could work outside cells, he suggested be called

"enzymes" (from Greek words meaning "in yeast"), because they resembled the ferments in yeast in their action.

Then, in 1897, the whole vitalist position was, in this
respect, unexpectedly exploded by the German chemist,
Eduard Buchner (1860–1917). He ground yeast cells with
sand until not one was left intact and then filtered the
ground-up material, obtaining a cell-free quantity of yeast
juice. It was his expectation that this juice would have
none of the fermenting ability of living yeast cells. It was
important, however, that the juice be kept from contamination with microorganisms or it would then contain
living cells after all and the test would not be a good one.

One time-tested method of preserving materials against
contamination by microorganisms is the addition of a
concentrated sugar solution. Buchner added this and
found, to his amazement, that the sugar began to undergo
a slow fermentation, although the mixture was absolutely
nonliving. He experimented further, killing yeast cells
with alcohol and finding that the dead cells would ferment sugar as readily as the live ones would.

As the nineteenth century drew to a close, it was recognized that all ferments, organized as well as unorganized,
were dead substances that might be isolated from cells
and made to do their work in the test tube. The name
"enzyme" was applied to all ferments alike and it was
therefore accepted that the cell contained no chemicals
that could work only in the presence of some life force.

Pasteur's dictum that without life there could be no
fermentation was found to apply only to situations as
they occurred in nature. The interfering hand of man
could so treat the yeast cell that though the cell and its
life was destroyed, the constituent enzymes remained intact and then fermentation could be made to proceed
without life.

This was the most serious defeat vitalism had yet en-
dured but, in a sense, the vitalist position was far from
shattered. Much remained to be discovered about the pro-

tein molecule (both enzymes and nonenzymes), and it could not be considered certain that the life force would not, at some point or other, make itself evident. In particular, Pasteur's (and Virchow's) other dictum that no cell could arise except from a pre-existing cell remained and, while that remained, there was *still* something special about life that perhaps the hand of mere man might not touch.

However, the heart went out of the vitalist position. Individual biologists might still speak diluted forms of vitalism in theory (and some do even today) but none seriously act upon it. It is generally accepted that life follows the laws that govern the inanimate world; that there is no problem in biology that is innately beyond solution in the laboratory, nor any life process that may not be imitated there in the absence of life.

The mechanistic view is supreme.

CHAPTER 9

The War Against Disease

Vaccination

In considering the great debates over evolution and vitalism, it is important to keep from forgetting that man's interest in biology as a science grew out of a preoccupation with medicine; with the disorders of the body. However far the science may fly off into the realm of theory and however serenely it may seem to hover beyond the concern of ordinary affairs of men, to that preoccupation it will return.

Nor is a concern with theory distracting or wasteful, for when, armed with an advance in theory, men turn to application of a science, how rapidly matters march. And although applied science may advance in a purely empirical fashion without theory, how slow and fumbling that is in comparison.

As an example, consider the history of infectious disease. Until nearly the dawn of the nineteenth century, doctors had been, by and large, helpless in the face of the vast plagues and epidemics that periodically swept across the land. And of the diseases that plagued mankind, one of the worst was smallpox. Not only did it spread like wildfire; not only did it kill one in three; but even those who survived were unfortunate, for their faces might easily be left so pitted and scarred that one could scarcely endure the sight of them.

One attack of smallpox, however, insured immunity to future attacks. For that reason, a very mild case of smallpox, leaving one virtually unscarred, was far, far better than none at all. In the former case, one was forever safe; in the latter, forever under the threat. In such places as Turkey and China, there were attempts, consequently, to catch the disease from those with mild cases. There was even deliberate inoculation with matter from the blisters produced by mild smallpox. The risk was terrible, for sometimes the disease, when caught, proved not mild at all in the new host.

In the early eighteenth century, such inoculation was introduced into England but did not really prove popular. However, the subject was in the air and under discussion and an English physician, Edward Jenner (1749–1823), began to consider the matter. There was an old-wives' tale in his native county of Gloucestershire to the effect that anyone who caught cowpox (a mild disease common to cattle that resembled smallpox in some ways) was thereafter immune not only to cowpox but to smallpox as well.

Jenner, after long and careful observation, decided to test this. On May 14, 1796, he found a milkmaid who had cowpox. He took the fluid from a blister on her hand and injected it into a boy who, of course, got cowpox in his turn. Two months later, he inoculated the boy again, not with cowpox, but with smallpox. It did not touch the youngster. In 1798, after repetition of the experiment, he published his findings. He coined the word "vaccination" to describe the technique. This is from the Latin word, "vaccinia," for cowpox, which, in turn, comes from the Latin word, "vacca," for cow.

Such was the dread of smallpox that for once an advance was greeted and accepted with almost no suspicion. Vaccination spread like wildfire over Europe and the disease was vanquished. Smallpox has never since been a major problem in any of the medically advanced nations. It was the first serious disease in the history of mankind to be so rapidly and completely brought under control.

But the advance could not be followed up in the absence of appropriate theory. No one as yet knew the cause of infectious disease (smallpox or any other), and the accident of the existence of a mild cousin of a major disease which could be used for inoculative purposes was not to happen again. Biologists simply had to learn to manufacture their own mild versions of a disease, and that required more knowledge than they possessed in Jenner's time.

The Germ Theory of Disease

The necessary theory came with Pasteur, whose interest in microorganisms dated from his concern with the fermentation problem (see page 91). This interest now led to something more.

In 1865, the silk industry in southern France was being dealt a staggering blow by a disease that was killing the silkworms, so the call went out once more for Pasteur. He

used his microscope and found a tiny parasite infesting
the sick silkworms and the mulberry leaves that were being fed to them. Pasteur's solution was drastic but rational. All infested worms and infected food must be
destroyed. A new beginning must be made with healthy
worms and clean food. This worked and the silk industry
was saved.

But to Pasteur it seemed that what was true of one
infectious disease might be true of others. A disease could
be caused by microorganisms. It could then be spread by
coughing, sneezing, or kissing, through wastes, through
contaminated food or water. In each case, the disease-
causing microorganism would spread from the sick man
to the healthy one. The physician in particular, thanks
to his necessary contact with the sick, might be a prime
agent of infection.

The last conclusion had indeed been reached by a Hungarian physician, Ignaz Philipp Semmelweiss (1818–65).
Without knowledge of Pasteur's theory he nevertheless
could not help but notice that the death rate from child-
bed fever among women in Vienna hospitals was dread-
fully high, while among women who gave birth at home
with the help of ignorant midwives it was quite low. It
seemed to Semmelweiss that doctors who went from the
dissecting room to the operating room must be carrying
the disease somehow. He insisted that doctors wash their
hands thoroughly before approaching the woman in la-
bor. Whenever he could carry that through, the death rate
fell. The offended doctors forced him out, however, and
the death rate rose again. Semmelweiss died defeated and
just too soon to see victory. (In the United States, at
about the same time, the American physician and poet,
Oliver Wendell Holmes (1809–94), carried on a similar
campaign against the dirty hands of obstetricians, and
won considerable abuse for himself.)

Once Pasteur advanced his "germ theory of disease,"
however, conditions slowly changed. There was now a

reason to wash, and however much conservative physicians might protest against the new-fangled notion, they were slowly forced into line. During the Franco-Prussian War, Pasteur managed to force doctors to boil their instruments before using them on wounded soldiers and to steam their bandages.

Meanwhile, in England, a surgeon, Joseph Lister (1827–1912), was doing his best to reform surgery. He was putting "anesthesia" into use, for instance. In this technique, a patient breathed a mixture of ether and air. This caused him to fall asleep and become insensible to pain. Teeth could be extracted, and operations performed, without torture. Several men had contributed to his discovery but the lion's share of the credit is usually given to an American dentist, William Thomas Green Morton (1819–68), who arranged to have a facial tumor removed from a patient under ether in the Massachusetts General Hospital in October 1846. This successful display of anesthesia quickly established it as part of surgical procedure.

However, Lister was distressed to find that even though an operation might be painless and successful, the patient might still die of the subsequent infection. When he heard of Pasteur's theory, the thought occurred to him that if the wound or surgical incision were sterilized, infection would not catch hold. He began by using carbolic acid (phenol) and found it worked. Lister had introduced "antiseptic surgery."

Gradually, less irritating and more effective chemicals were found for the purpose. Surgeons took to wearing sterilized rubber gloves and face masks. Surgery was finally made safe for mankind. If Pasteur's germ theory had done this alone, it might have been enough to make it the most important single discovery in the history of medicine. However, it accomplished more, much more, and its unparalleled importance cannot be challenged.

Bacteriology

One couldn't expect to keep all deadly microorganisms away from all human beings at all times. Sooner or later, exposure to disease was certain. What then?

To be sure, the body had ways of fighting microorganisms, since it could recover from infections spontaneously. In 1884, the Russian-French biologist, Ilya Ilitch Mechnikov (1845–1916), was to find a dramatic example of such "counterbacterial warfare." He was able to show that the white corpuscles of the blood, equipped with the capacity to leave the blood vessels if necessary, flocked to the site of infections or of bacterial invasion. What followed was very much like a pitched battle between bacteria and white corpuscles, with the latter not necessarily always winning, but winning often enough to do a great deal of good.

Yet there had to be more subtle antibacterial weapons, too, since in the case of many diseases, recovery from one attack meant immunity thereafter, although no visible changes in the body could be found. A logical explanation for this was that the body had developed some molecule (an "antibody") which could be used to kill an invading microorganism or neutralize its effect. This would explain the effect of vaccination, since the body might have developed an antibody against the cowpox microorganism and found it usable against the very similar smallpox microorganism.

Now at last that victory could be repeated not through an attack on the disease itself but on the microorganism that caused the disease. Pasteur showed the way in connection with anthrax, a deadly disease that ravaged herds of domestic animals. Pasteur searched for a microorganism that would cause the disease and found it in the form of a particular bacterium. He heated a preparation of such bacteria just long enough to destroy their ability to bring on the disease. These helpless "attenuated bacteria," by their mere presence would force a body to develop antibodies against them, antibodies that could be used against the fresh, deadly bacteria, too.

In 1881, Pasteur carried through a most dramatic experiment. Some sheep were inoculated with his attenuated bacteria while other sheep were not. After a time, all the sheep were exposed to deadly anthrax bacteria. Every sheep that had been first inoculated survived without ill-effect; the others caught anthrax and died.

Similar methods were established by Pasteur in the fight against chicken cholera and, most dramatically of all, against rabies (or hydrophobia), the disease caused by the bite of a "mad dog." In effect, he was creating artificial cowpoxes, so to speak, to protect men and animals against a whole variety of smallpoxes.

The success of Pasteur's germ theory created an intense new interest in bacteria. The German botanist, Ferdinand Julius Cohn (1828–98), had been interested in microscopic plant cells in his youth. He showed, for instance, that plant protoplasm was essentially identical with animal protoplasm. In the 1860s, however, he turned to bacteria and, in 1872, published a three-volume treatise on the little creatures in which the first systematic attempt was made to classify them into genera and species. For that reason, Cohn may be considered the founder of modern *bacteriology*.

Cohn's most important discovery, however, was of a young German doctor named Robert Koch (1843–1910). In 1876, Koch had isolated the bacterium causing anthrax and learned to cultivate it (as Pasteur was doing in France). Koch brought his work to Cohn's attention, and the enthusiastic Cohn sponsored him vigorously.

Koch learned to grow bacteria on a solid gel, such as gelatin (for which, later, agar-agar, a product of seaweed, was substituted), instead of in liquid. This made a great deal of difference. In liquid, bacteria of different varieties mix easily and it is difficult to tell which variety may be causing a particular disease.

If, however, a culture were smeared on a solid medium, an isolated bacterium would divide and redivide, producing many new cells that would not be able to move from the spot. Though the original culture might be a mixture of many species of bacteria, that one solid colony would have to be a pure variety. If it produced a disease, there could be no question as to which variety was responsible.

Originally, Koch placed his gel on a flat piece of glass, but an assistant, Julius Richard Petri (1852–1921), substituted a shallow dish with a glass cover. Such "Petri dishes" have been used in bacteriology ever since.

Working with pure cultures, Koch was able to evolve rules for the detection of the microorganism causing a particular disease. He and his assistants discovered many such, and the high point in Koch's career was his identification, in 1882, of the bacterium that caused tuberculosis.

Insects

Bacteria need not be the only causative agents of an infectious disease and that is why Pasteur's discovery is called the "germ theory," "germ" signifying microorganisms generally and not bacteria only. In 1880, for instance, a French physician, Charles Louis Alphonse Laveran (1845–1922), while stationed in Algeria, found the causative agent of malaria. This was particularly exciting in itself since malaria is a widespread disease over most of the tropical and subtropical world, killing more human beings all told than any other. What made the discovery particularly interesting, however, was that the agent was not a bacterium, but a protozoon, a one-celled animal.

Indeed an illness might not even be caused by a microorganism. In the 1860s, a German zoologist, Karl Georg Friedrich Rudolph Leuckart (1822–98), in his studies of invertebrates, found himself particularly interested in those which lived parasitically within the bodies of other organisms; thus founding the science of *parasitology*. He found that all the invertebrate phyla had their parasitic representatives. A number of these infest men, and such creatures as flukes, hookworms, and tapeworms—far from microscopic—can produce serious illness.

What's more, a multicellular animal, even if not the direct causative agent of a disease, may nevertheless be the carrier of infection, which is just as bad. Malaria was the first disease in which this aspect of infection became important. An English physician, Ronald Ross (1857–1932), investigated suggestions that perhaps mosquitoes spread malaria from person to person. He collected and dissected mosquitoes and, in 1897, finally located the malarial parasite in the anopheles mosquito.

This was a most useful discovery, since the mosquito represented a weak point in the chain of infection. It could be easily shown that malaria did not spread by direct contact (the parasite, it seems, must pass through a life stage in the mosquito before it can enter man again), so why not simply do away with the mosquito? Why not sleep under mosquito netting? Why not drain swamps in which mosquitoes breed? This worked, and where such methods were used, the incidence of malaria declined.

Another deadly disease, one that during the eighteenth and nineteenth centuries periodically ravaged the east coast of the United States, was yellow fever. During the Spanish-American War, the American Government grew particularly disease-conscious, since germs killed far more American soldiers in Cuba than Spanish guns did. In 1899, after the war, an American military surgeon, Walter Reed (1851–1902), was sent to Cuba to see what could be done.

He found that yellow fever was not spread by direct contact and, in view of Ross's work, he suspected mosquitoes, this time another species, the Aedes mosquito.

Doctors working with Reed allowed themselves to be bitten by mosquitoes that had been biting infected men, and some of them got the disease. One young doctor, Jesse William Lazear (1866–1900), died as a result, a true martyr to the cause of humanity. The case was proved.

Another American army surgeon, William Crawford Gorgas (1854–1920), used mosquito-fighting methods to wipe out yellow fever in Havana, and was then assigned to Panama. The United States was trying to build a canal there, although France had failed in a previous attempt. The engineering difficulties were great, to be sure, but it was the high death rate from yellow fever that really blunted all efforts. Gorgas brought the mosquito under control, stopped the disease cold, and in 1914, the Panama Canal was opened.

Nor was the mosquito the only insect that played the role of villain. In 1902, a French physician, Charles Jean Henri Nicolle (1866–1936), was appointed director of the Pasteur Institute in Tunis, North Africa. There, he had occasion to study the dangerous and highly infectious disease, typhus fever.

Nicolle noticed that while outside the hospital the disease was extremely contagious, it was not contagious at all within the hospital. Patients in the hospital were stripped of their clothes and scrubbed down with soap and water on admission, and it occurred to Nicolle that the infective agent must be something in the clothing, something that could be removed from the body by washing. His suspicion fell on the body louse, and, through animal experiments, he proved his case by showing that only through the bite of the louse could the disease be transmitted. Similarly, in 1906, the American pathologist, Howard Taylor Ricketts (1871–1911), showed that Rocky Mountain spotted fever was transmitted by the bite of cattle ticks.

Food Factors

The germ theory dominated the minds of most physicians through the last third of the nineteenth century but there were a few who resisted it. The German pathologist, Virchow (see page 69), was the most eminent of these. He preferred to think of disease as being caused by some irritation from within, rather than some agent from without. He was also a man of strong social consciousness who spent some decades in Berlin city politics and in the national legislature. He pushed through important improvements in such matters as a purified water supply and an efficient sewage system. Pettenkofer (see page 89) was another of this type and he and Virchow were among the founders of modern notions of *public hygiene* (the study of the prevention of disease in the community).

Such improvements interfered with the easy transmission of disease (whether Virchow believed in germs or not) and were probably as instrumental in putting an end to the epidemics that had, until the mid-nineteenth century, plagued Europe, as was the more direct concern with germs themselves.

If Hippocrates' interest in cleanliness retained its force in the days of germ-consciousness, that was to be expected. Perhaps more surprising was the fact that Hippocrates' advice as to a good and varied diet also retained its force, and not only for the sake of general well-being, but as a specific method of preventing specific diseases. Poor diet as a cause of disease seemed to many, during the germ-conscious generation from 1870 to 1900, to be an outmoded notion, and yet there was strong evidence to show that it was not at all outmoded.

Thus, in the early days of the Age of Exploration, men spent long months on board ship, living only on food items that could keep over those periods, since refrigeration was unknown. In those days, scurvy was the dreaded disease of seamen. A Scottish physician, James Lind (1716–94), took note of the fact that scurvy accompanied monotonous diet not only on shipboard, but in besieged cities and in prisons. Could a missing dietary item be the cause of the disease then?

In 1747, Lind tried different food items on scurvy-ridden sailors and found that citrus fruits worked amazingly well in effecting relief. Slowly, this device was adopted. Captain James Cook (1728–79), the great English explorer, fed citrus fruit to his men on his Pacific voyages in the 1770s and lost only one man to scurvy. In 1795, the British Navy, under the pressures of a desperate war with France, began compulsory feeding of lime juice to sailors, and scurvy was wiped out on British ships.

However, such empirical progress is slow in the absence of the necessary advances in basic science. Through the nineteenth century, the major discoveries in nutrition concerned the importance of protein and, in particular, the fact that some proteins were "complete" and could support life when present in the diet, while others, like gelatin, were "incomplete" and could not (see page 86).

An explanation for this difference among proteins came only when the nature of the protein molecule was better understood. In 1820, the complex molecule of gelatin was broken down by treatment with acid and a simple molecule, named "glycine," was isolated. Glycine belonged to a class of compounds called "amino acids."

At first it was assumed that glycine was *the* building block of proteins, as the simple sugar, glucose, was the building block of starch. However, as the nineteenth century progressed, this theory turned out to be inadequate. Other simple molecules were obtained out of various proteins. All were of the class, amino acid, but they differed in detail. Protein molecules were not built out of one, but out of a number of amino acids. By 1900, a dozen different amino-acid building blocks were known.

It was quite possible, then, that proteins might differ

in the relative proportions of the different amino acids they contained. A particular protein might even be lacking altogether in one or more particular amino acids and those amino acids might be essential to life.

The first to show that this was indeed so was an English biochemist, Frederick Gowland Hopkins (1861–1947). In 1900, he had discovered a new amino acid, tryptophan, and had developed a chemical test that would indicate its presence. Zein, a protein isolated from corn, did not respond to that test and therefore lacked tryptophan. Zein was an incomplete protein and would not support life where it was the sole protein in the diet. If, however, a bit of tryptophan was added to zein, the life of the experimental animals was prolonged.

Similar experiments conducted during the early decades of the twentieth century made it quite clear that some amino acids could be formed by the mammalian body from substances usually available in the tissues. A few, however, could not so be manufactured and had to be present, intact, in the diet. It was the absence of one or more of these "essential amino acids" that made some proteins incomplete and brought on sickness and eventual death.

Thus was introduced the concept of a "food factor": any compound that could not be made in the body, and that had to be present in the diet, intact, if life was to be maintained. To be sure, amino acids were not serious medical problems, however interesting they might be to nutritionists. An amino acid deficiency was generally brought on by artificial and deliberately lopsided diets. A natural diet, even a poor one, usually supplied enough of each amino acid.

If a disease such as scurvy could be cured by lime juice, it was reasonable to suppose that the lime juice was supplying a missing food factor. It was not likely however that the food factor was an amino acid. In fact, all the constituents of lime juice known to the nineteenth-century biologist would not, taken singly or together, cure scurvy. The food factor involved must therefore be a substance that was necessary only in trace quantities and one that might well be quite different, chemically, from the usual components of food.

Actually, the mystery was not as hard to solve as it might seem. Even as the concept of the essential amino acid was worked out, other more subtle food factors, required only in traces, were also being discovered, and, as it happened, not through a study of scurvy.

Vitamins

A Dutch physician, Christiaan Eijkman (1858–1930), was sent to Java in 1886 to study the disease beriberi. There was reason to think that the disease might be the result of imperfect diet. Japanese sailors had suffered from it extensively—then ceased suffering in the 1880s when a Japanese admiral added milk and meat to a diet that, previously, had been almost exclusively fish and rice.

Eijkman, however, was immersed in germ theory and was sure beriberi was a bacterial disease. He brought chickens with him and hoped to cultivate the germ in them. In this he failed. However, during the course of 1896, his chickens came down spontaneously with a disease very much like beriberi. Before Eijkman could do much about it, the disease vanished.

Searching for causes, Eijkman found that for a certain period of time the chickens had been fed on polished rice from the hospital stores and it was then they sickened. Put back on commercial chicken feed they recovered. Eijkman found further that he could produce the disease at will and cure it, too, by simply changing the diet.

Eijkman did not appreciate the true meaning of this at first. He thought there was a toxin of some sort in rice grains and that this was neutralized by something in the hulls. The hulls were removed when rice was polished, leaving the toxin in the polished rice unneutralized (so Eijkman thought).

However, why assume the presence of two different unknown substances, a toxin and an antitoxin, when it was only necessary to assume one: some food factor required in traces? The outstanding exponents of this latter view were Hopkins himself (see page 109) and a Polish-born biochemist, Casimir Funk (1884–). Each suggested that not only beriberi, but also such diseases as scurvy, pellagra, and rickets were caused by the absence of trace food factors.

Under the impression that these food factors belonged to the class of compounds known as "amines," Funk suggested, in 1912, that these factors be named "vitamines" ("life amines"). The name was adopted, but since it turned out that the factors were not all amines, the name was changed to "vitamins."

The Hopkins-Funk "vitamin hypothesis" was borne out in full, and the first third of the twentieth century saw a variety of diseases overcome wherever sensible dietary rules could be established. As an example, the Austrian-American physician, Joseph Goldberger (1874–1929), showed, in 1915, that the disease, pellagra, endemic in the American south, was caused by no germ. Instead, it was due to the lack of a vitamin and it could be abolished if milk were added to the diet of those who suffered from it.

At first, nothing was known about the vitamins other than their ability to prevent and to cure certain diseases. The American biochemist, Elmer Vernon McCollum (1879–), introduced, in 1913, the device of referring to them by letters of the alphabet, so that there was vitamin A, vitamin B, vitamin C, and vitamin D. Eventually, vitamins E and K were added. It turned out that food containing vitamin B actually contained more than one factor capable of correcting more than one set of symptoms. Biologists began to speak of vitamin B_1, vitamin B_2 and so on.

It was the absence of vitamin B_1 that brought on beri-beri, and the absence of vitamin B_6 that caused pellagra. The absence of vitamin C led to scurvy (and it was the presence of vitamin C in small amounts in citrus fruits that had enabled Lind to cure scurvy) and the absence of vitamin D brought on rickets. The absence of vitamin A affected vision and caused night blindness. These were the major vitamin-deficiency diseases and as knowledge of vitamins increased, they ceased to be a serious medical problem.

CHAPTER 10

The Nervous System

Hypnotism

Another variety of illnesses that certainly did not come under Pasteur's germ theory was the mental diseases. These had confused, frightened, and overawed mankind from earliest times. Hippocrates approached them in a rationalistic fashion (see page 4), but the vast majority of mankind maintained the superstitious view. No doubt the feeling that madmen were under the control of demons helped explain the fearful cruelty with which the mentally diseased were treated up to the nineteenth century.

The first breath of a new attitude in this respect came with a French physician, Philippe Pinel (1745–1826). He considered insanity a mental illness and not demonic

possession, and published his views on what he called
"mental alienation." In 1793, with the French Revolution
in full swing and with the smell of change in the air, Pinel
was placed in charge of an insane asylum. There he struck
off the chains from the inmates and for the first time
allowed them to be treated as sick human beings and not
as wild animals. The new view spread outward only
slowly, however.

Even when a mental disorder was not serious enough
to warrant hospitalization, it might still give rise to un-
pleasant and very real physical symptoms ("hysteria" or
"psychosomatic illness"). Such symptoms, originating in
a mental disorder, might be relieved by a course of treatment that affected the mind. In particular, if a person
believes that a treatment will help him, that treatment
may indeed help him in so far as his ailment is psychosomatic. For this reason, exorcism, whether that of the priest
or the witch doctor, can be effective.

Exorcism was brought from theology into biology by an
Austrian physician, Friedrich Anton Mesmer (1733–
1815), who used magnets for his treatments at first. He
abandoned these and made passes with his hands, utiliz-
ing what he called "animal magnetism." Undoubtedly,
he effected cures.

Mesmer found that his cures were more rapid if he
placed the patient into a trancelike condition by having
him fix his attention on some monotonous stimulus. By
this procedure (sometimes called "mesmerism," even today), the patient's mind was freed of bombardment by
the many outside stimuli of the environment and was
concentrated on the therapist. The patient became, therefore, more "suggestible."

Mesmer was a great success for a time, particularly in
Paris, where he arrived in 1778. However, he overlarded
his techniques with a mysticism that verged on charla-
tanry and, furthermore, he attempted to cure diseases that
were not psychosomatic. These diseases he did not cure,

of course, and the patients, as well as competing physicians using more orthodox methods, complained. A commission of experts was appointed to investigate him and they turned in an unfavorable report. Mesmer was forced to leave Paris and retire to Switzerland and obscurity.

Yet the value of the essence of his method remained. A half-century later, a Scottish surgeon, James Braid (1795–1860), began a systematic study of mesmerism, which he renamed "hypnotism" (from a Greek word meaning "sleep"). He reported on it in a rationalistic manner in 1842, and the technique entered medical practice. A new medical specialty, *psychiatry*, the study and treatment of mental disease, came into being.

This specialty gained real stature with an Austrian physician, Sigmund Freud (1856–1939). During his medical-school days and for a few years thereafter, Freud was engaged in orthodox research on the nervous system. He was the first, for instance, to study the ability of cocaine to deaden nerve endings. Carl Koller (1857–1944), an interne at the hospital in which Freud was working, followed up that report and, in 1884, used it successfully during an eye operation. This was the first use of a "local anesthetic," that is, one which would deaden a specific area of the body, making it unnecessary to induce over-all insensibility for a localized operation.

In 1885, Freud traveled to Paris, where he was introduced to the technique of hypnotism and where he grew interested in the treatment of psychosomatic illness. Back in Vienna, Freud began to develop the method further. It seemed to him that the mind contained both a conscious and an unconscious level. Painful memories, or wishes and desires of which a person was ashamed, might, he felt, be "repressed"; that is, stored in the unconscious mind. The person would not consciously be aware of this store, but it would be capable of affecting his attitudes and actions and of producing physical symptoms of one sort or another.

Under hypnotism, the unconscious mind was apparently tapped, for the patient could bring up subjects that, in the normally conscious state, were blanks. In the 1890s, however, Freud abandoned hypnotism in favor of "free association," allowing the patient to talk randomly and freely, with a minimum of guidance. In this fashion, the patient was gradually put off-guard, and matters were revealed which, in ordinary circumstances, would be carefully kept secret even from the patient's own conscious mind. The advantage of this over hypnotism lay in the fact that the patient was at all times aware of what was going on and did not have to be informed afterward of what he had said.

Ideally, once the contents of the unconscious mind were revealed, the patient's reactions would no longer be unmotivated to himself, and he would be more able to change those reactions through an understanding of his now-revealed motives. This slow analysis of the contents of the mind was called "psychoanalysis."

To Freud, dreams were highly significant, for it seemed to him that they gave away the contents of the unconscious mind (though usually in a highly symbolized form) in a manner that was not possible during wakefulness. His book *The Interpretation of Dreams* was published in 1900. He further felt that the sexual drive, in its various aspects, was the most important source of motivation, even among children. This last view roused considerable hostility on the part of the public as well as of much of the medical profession.

Beginning in 1902, a group of young men had begun to gather about Freud. They did not always see eye to eye with him and Freud was rather unbending in his views and not given to compromise. Men such as the Austrian psychiatrist, Alfred Adler (1870–1937), and the Swiss psychiatrist, Carl Gustav Jung (1875–1961), broke away and established systems of their own.

The Nerves and Brain

The vast complexity of the human mind is such, however, that belief in psychiatry remains very largely a matter of personal opinion. The different schools maintain their own views and there are few objective ways of deciding among them. If further advance is to be made, it will come when the basic science of the nervous system (*neurology*) is sufficiently developed.

Neurology began with a Swiss physiologist, Albrecht von Haller (1708–77), who published an eight-volume textbook on human physiology in the 1760s. Before his time it had been generally accepted that the nerves were hollow and carried a mysterious "spirit" or fluid, much as veins carried blood. Haller, however, discarded this and reinterpreted nerve action on the basis of experiment.

For instance, he recognized that muscles were "irritable"; that is, that a slight stimulus of a muscle would produce a sharp contraction. He also showed, however, that a slight stimulus to a nerve would produce a sharp contraction in the muscle to which it was attached. The nerve was the more irritable of the two and Haller judged that it was nervous stimulation rather than direct muscular stimulation that controlled the movements of muscles.

Haller also showed that tissues themselves do not experience a sensation but that the nerves channel and carry the impulses that produce the sensation. Furthermore, he showed that nerves all lead to the brain or the spinal cord, which are thus clearly indicated to be the centers of sense perception and responsive action. He experimented by stimulating or damaging various parts of the animal brain and then noting the type of action or paralysis that resulted.

Haller's work was carried further by the German physician, Franz Joseph Gall (1758–1828), who began lecturing on the subject in 1796. He showed that the nerves led not merely to the brain, but to the "gray matter" on the surface of the brain. The "white matter," below the surface, he held to be connective substance.

Like Haller, Gall felt that particular parts of the brain were in control of particular parts of the body. He carried this to extremes, feeling that specific parts of the brain were assigned not only to particular sense perceptions and to particular muscle movements, but also to all sorts of emotional and temperamental qualities. This view was carried to the point of absurdity by his later followers who felt that these qualities could be detected, when present in excess, by feeling bumps on the skull. Thus was developed the pseudoscience of "phrenology."

The silliness of phrenology obscured the fact that Gall was partly right and that the brain did indeed have specialized areas. This possibility was lifted out of pseudoscience and back to rational investigation by the French brain surgeon, Paul Broca (see page 69). As a result of a number of post mortems, he showed, in 1861, that patients, suffering from a loss of the ability to speak, possessed damage to a certain specific spot on the upper division of the brain, the cerebrum. The spot was on the third convolution of the left frontal lobe which is still called "Broca's convolution."

By 1870, two German neurologists, Gustav Theodor Fritsch (1838–91) and Eduard Hitzig (1838–1907), had gone even further. They exposed the brain of a living dog and stimulated various portions with an electric needle. They found that the stimulation of a particular spot would induce a particular muscular movement and in this way, they could map the body, so to speak, on the brain. They were able to show that the left cerebral hemisphere controlled the right part of the body while the right cerebral hemisphere controlled the left.

Thus, there came to be no doubt that not only did the brain control the body, but that it did so in a highly specific way. It began to seem that there was at least a conceivable chance that all mental function could be related in one way or another to brain physiology. This would make the mind merely an extension of the body and threatened to bring man's highest powers within the mechanistic domain.

More fundamentally still, the cell theory, when it came into being, was eventually applied to the nervous system. The biologists of the mid-nineteenth century had detected nerve cells in the brain and spinal cord, but were vague as to the nature of the nerve fibers themselves. It was the German anatomist, Wilhelm von Waldeyer (1836–1921), who clarified the matter. He maintained, in 1891, that the fibers represented delicate extensions from the nerve cells and formed an integral part of them. The whole nervous system, therefore, consisted of "neurons"; that is, of nerve cells proper, plus their extensions. This is the "neuron theory." Furthermore, Waldeyer showed that extensions of different cells might approach closely but did not actually meet. The gaps between neurons later came to be called "synapses."

The neuron theory was placed on a firm footing through the work of the Italian cytologist, Camillo Golgi (1844–1926), and the Spanish neurologist, Santiago Ramón y Cajal (1852–1934). In 1873, Golgi developed a cell stain consisting of silver salts. By use of this material, he revealed structures within the cell ("Golgi bodies") whose functions are still unknown.

Golgi applied his staining method to nerve tissue in particular and found it well adapted for the purpose. He was able to see details not visible before, to make out the fine processes of the nerve cells in unprecedented detail, and to show synapses clearly. Nevertheless, he opposed Waldeyer's neuron theory when that was announced.

Ramón y Cajal, however, upheld the neuron theory strongly. Using an improved version of the Golgi staining technique he demonstrated details that established the neuron theory beyond question and worked out the cellular structure of the brain and spinal cord, and of the retina of the eye, too.

Behavior

The neuron theory could be applied usefully to the problem of animal behavior. As early as 1730, Stephen Hales (see pp. 46–47), found that if he decapitated a frog, it would still kick its leg if its skin were pricked. Here a body reacted mechanically without the aid of the brain. This initiated a study of the more or less automatic "reflex action," where a response follows hard upon a stimulus, according to a set pattern and without interference of the will.

Even the human being is not free of such automatic action. A blow just beneath the kneecap will produce the familiar knee jerk. If one's hand comes casually into contact with a hot object, it is snatched away at once, even before one becomes consciously aware that the object is hot.

The English physiologist, Charles Scott Sherrington (1861–1952), studied reflex action and founded *neurophysiology*, as Golgi and his stain had earlier founded *neuroanatomy*. Sherrington demonstrated the existence of the "reflex arc," a complex of at least two, and often more than two, neurons. Some sense impression at one place sent a nerve impulse along one neuron, then over a synapse (Sherrington invented the word), then, via a returning neuron, back to another place, where it stimulated muscle action or, perhaps, gland secretion. The fact that there might be one or more intermediate neurons between the first and last did not affect the principle.

It could seem that synapses were so arranged that some were crossed by the impulse more easily than others. Thus, there might be particular "pathways" that were easily traveled among the interlacing cobweb of neurons that made up the nervous system.

It could further be supposed that one pathway might open the way for another; that, in other words, the response of one reflex action might act as the stimulus for a second which would produce a new response acting as a stimulus for a third and so on. A whole battery of reflexes might then make up the more-or-less complex behavior pattern we call an "instinct."

A relatively small and simple organism like an insect could be very little more than a bundle of instincts. Since the "nerve pathways" can be conceived of, easily enough, as being inherited, one can understand that instincts are inherited and are present from birth. Thus, a spider can spin a web perfectly, even if it has never seen a web being spun; and each species of spider will spin its own variety of web.

Mammals (and man in particular) are relatively poor in instincts but are capable of learning, that is, of evolving new behavior patterns on the basis of experience. Even though the systematic study of such behavior in terms of the neuron theory may be difficult, it is possible to analyze behavior in a purely empirical fashion. Throughout history, intelligent men have learned to calculate how human beings would react under particular circumstances and this ability has made them successful leaders of men.

The application of quantitative measurement to the human mind, however (at least to its ability to sense the environment), begins with the German physiologist, Ernst Heinrich Weber (1795–1878). In the 1830s, he found that the size of the difference between two sensations of the same kind depended on the logarithm of the intensity of the sensations.

Just as in lighting a room, if we begin with a room lit by one candle, a second equal candle is sensed as brightening the room by an amount we call x. Further brightenings of that degree are not produced by single additional candles but by larger and larger sets of candles. First one additional candle will suffice to brighten the room by x, then two more candles will be required for a further brightening by x, then four more, then eight more, and so on. This rule was popularized in 1860 by the German physicist, Gustav Theodor Fechner (1801–87), and is sometimes called the "Weber-Fechner law" in consequence. This initiated *psychophysics*, the quantitative study of sensation.

The study of behavior generally (*psychology*) is less easily reduced to mathematics, but it can be made experimental. The founder of this approach was the German physiologist, Wilhelm Wundt (1832–1920), who set up the first laboratory dedicated to experimental psychology in 1879. Out of his work arose patterns of experimentation which involved setting rats to solving mazes and chimpanzees to reasoning out methods for reaching bananas. This was applied to human beings, too, and in fact the asking of questions and setting of problems was used in the attempted measurement of human intelligence. The French psychologist, Alfred Binet (1857–1911), published his first IQ (intelligence quotient) tests in 1905.

More fundamental studies, relating behavior more directly to the nervous system, were made by the Russian physiologist, Ivan Petrovich Pavlov (1849–1936). In the earlier portion of his career, he was interested in the nerve control of the secretion of digestive juices. With the turn of the century, however, he began to study reflexes.

A hungry dog which is shown food will salivate. This is a reasonable reflex, for saliva is needed for the lubrication and digestion of food. If a bell is made to ring every time the dog is shown food, it will associate the sound of the bell with the sight of food. Eventually, it will salivate as soon as it hears the sound of the bell, even though it sees no food. This is a "conditioned reflex." Pavlov was able to show that all sorts of reflexes could be set up in this fashion.

A school of psychology, "behaviorism," grew up which

maintained that all learning was a matter of the development of conditioned reflexes and of new hookups, so to speak, of the nerve network. One related the appearance of the print patterned "chair," with the sound pattern produced in pronouncing the word, and with the actual object in which one sits, until finally the mere sight of "chair" induces the thought of the object at once. The outstanding exponents of this school at its extreme were two American psychologists, John Broadus Watson (1878–1958) and, later, Burrhus Frederic Skinner (1904–   ).

Behaviorism is an extremely mechanistic view of psychology, and reduces all phases of the mind to the physical pattern of a complex nerve network. However, the current feeling is that this is too simple an interpretation. If the mind is to be interpreted mechanistically, it must be done in more subtle and sophisticated fashion.

Nerve Potentials

When considering a nerve network, it is easy to talk about impulses traveling along various pathways through the network, but of what, exactly, do those impulses consist? The ancient doctrine of a "spirit" flowing through the nerves had been smashed by Haller and Gall; but it arose again almost at once, albeit in a new form, when the Italian anatomist, Luigi Galvani (1737–98), discovered, in 1791, that the muscles of a dissected frog could be made to twitch under electrical stimulation. He declared there was such a thing as "animal electricity" produced by muscle.

This suggestion, in its original form, was not correct, but properly modified it proved fruitful. The German physiologist, Emil Du Bois-Reymond (1818–96), wrote a paper on electric fishes while still a student, and this initiated in him a lifelong interest in the electrical phenomena within tissues. Beginning in 1840, he set about refining old instruments and inventing new ones, instruments with which he might detect the passage of tiny currents in nerve and muscle. He was able to show that the nerve impulse was accompanied by a change in the electrical condition of the nerve. The nerve impulse was, in part at least, electrical in nature, and certainly electricity was as subtle a fluid as the old believers in a nervous "spirit" could have wished.

Electrical changes not only moved along the nerve but along muscles as well. In the case of a muscle undergoing rhythmic contractions, as was true of the heart, the electric changes were also rhythmic. In 1903, the Dutch physiologist, Willem Einthoven (1860–1927), devised a very sensitive "string galvanometer" capable of detecting extremely faint currents. He used it to record the rhythmically changing electric potentials of the heart through electrodes placed on the skin. By 1906, he was correlating the "electrocardiograms" (EKG) which he was recording, with various types of heart disorders.

A similar feat was performed in 1929 by the German psychiatrist, Hans Berger (1873–1941), who attached electrodes to the skull and recorded the rhythmically changing potentials that accompany brain activity. The "electroencephalograms" (EEG) are extremely complicated and hard to interpret. However, there are easily noted changes where extensive brain damage exists, as when tumors are present. Also the old "sacred disease" of epilepsy (see page 5) reveals itself in the form of changes in the EEG.

Electric potentials cannot, however, be the entire answer. An electrical impulse traveling along a nerve ending cannot, of itself, cross the synaptic gap between two neurons. Something else has to cross and initiate a new electrical impulse in the next neuron. The German physiologist, Otto Loewi (1873–1961), demonstrated, in 1921, that the nerve impulse involved a chemical change as well as an electrical one. A chemical substance, set free by the stimulated nerve, crossed the synaptic gap. The particular chemical was quickly identified by the English physiologist, Henry Hallet Dale (1875-   ), to be a compound called "acetylcholine."

Other chemicals have since been discovered to be related to nerve action in one fashion or another. Some have been found which will produce the symptoms of mental disorders. Such *neurochemistry* is as yet in its infancy, but it is hoped that it will eventually represent a powerful new means of studying the human mind.

CHAPTER 11

Blood

Hormones

The success of the neuron theory was, like that of the germ theory, not absolute. It did not carry quite all before it. The electrical messengers coursing along the nerve were not the only controls of the body. There were chemical messengers, too, making their way through the blood stream.

In 1902, for instance, two English physiologists, Ernest Henry Starling (1866–1927) and William Maddock Bayliss (1866–1924), found that even when all the nerves to the pancreas (a large digestive gland) were cut, it still performed on cue; it secreted its digestive juice as soon as the acid food contents of the stomach entered the intestine. It turned out that the lining of the small intestine, under the influence of the stomach acid, secreted a substance which Starling and Bayliss named "secretin." It was this secretin that stimulated the pancreatic flow.

Two years later, Starling suggested a name for all substances discharged into the blood by a particular "endo-
crine gland" for the purpose of rousing some other organ
or organs to activity. The word was "hormone" from
Greek words meaning "to rouse to activity."

The hormone theory proved extraordinarily fruitful, for
it was found that a large number of hormones, washing
through the blood in trace concentrations, interlaced their
effects delicately to maintain a careful balance among the
chemical reactions of the body, or to bring about a well-
controlled change where change was necessary. Already,
the Japanese-American chemist, Jokichi Takamine (1854–
1922), had, in 1901, isolated a substance from the adrenal
glands which is now called epinephrine (or Adrenalin—
a trade name) and this was eventually recognized as a
hormone. It was the first hormone to be isolated and to
have its structure determined.

One process that was quickly suspected of being hor-
mone-controlled was that of the basal metabolic rate.
Magnus-Levy had shown the connection between changes
in BMR and thyroid disease (see pp. 89–90), and the
American biochemist, Edward Calvin Kendall (1886–   ),
was able, in 1916, to isolate a substance from the thyroid
gland, which he called "thyroxine." This proved, indeed,
to be a hormone whose production in small quantities
controlled the BMR of the body.

The most spectacular early result of hormone work,
however, was in connection with the disease, diabetes
mellitus. This involved a disorder in the manner in which
the body broke down sugar for energy, so that a diabetic
accumulated sugar in his blood to abnormally high levels.
Eventually, the body was forced to get rid of the excess
sugar through the urine, and the appearance of sugar in
the urine was symptomatic of an advanced stage of the
disease. Until the twentieth century, the disease was certain death.

Suspicion arose that the pancreas was somehow connected with the disease, for in 1893, two German physiologists, Joseph von Mering (1849–1908) and Oscar Minkowski (1858–1931), had excised the pancreas of experimental animals and found that severe diabetes developed quickly. Once the hormone concept had been propounded by Starling and Bayliss, it seemed logical to suppose that the pancreas produced a hormone which controlled the manner in which the body broke down sugar.

Attempts to isolate the hormone from the pancreas, as Kendall had isolated thyroxine from the thyroid gland failed, however. Of course, the chief function of the pancreas was to produce digestive juices, so that it had a large content of protein-splitting enzymes. If the hormone were itself a protein (as, eventually, it was found to be) it would break down in the very process of extraction.

In 1920, a young Canadian physician, Frederick Grant Banting (1891–1941), conceived the notion of tying off the duct of the pancreas in the living animal and then leaving the gland in position for some time. The digestive-juice apparatus of the gland would degenerate, since no juice could be delivered; while those portions secreting the hormone directly into the blood stream would (he hoped) remain effective. In 1921, he obtained some laboratory space at the University of Toronto and with an assistant, Charles Herbert Best (1899–   ), he put his notion into practice. He succeeded famously and isolated the hormone "insulin." The use of insulin has brought diabetes under control, and while a diabetic cannot be truly cured even so and must needs submit to tedious treatment for all his life, that life is at least a reasonably normal and prolonged one.

Thereafter, other hormones were isolated. From the ovaries and testicles, the "sex hormones" (controlling the development of secondary sexual characteristics at puberty, and the sexual rhythm in females) were isolated by the German chemist Adolph Friedrich Johannes Butenandt (1903–   ), in 1929 and the years thereafter.

Men such as Kendall, the discoverer of thyroxine, and the Polish-Swiss chemist, Tadeus Reichstein (1897–), isolated a whole family of hormones, the "corticoids," from the outer portions (or "cortex") of the adrenal glands. In 1948, one of Kendall's associates, Philip Showalter Hench (1896–), was able to show that one of these corticoids, "cortisone," had a beneficial effect on rheumatoid arthritis.

The pituitary gland, a small structure at the base of the brain, was shown, in 1924, by the Argentinian physiologist, Bernardo Alberto Houssay (1887–), to be involved somehow with sugar breakdown. It turned out, later on, to have other important functions as well. The Chinese-American biochemist, Cho Hao Li (1913–), in the 1930s and 1940s, isolated a number of different hormones from the gland. One, for instance, is "growth hormone," which controls the over-all rate of growth. When produced in excessive amounts, a giant results; in deficient amounts, a midget is produced.

The study of hormones, *endocrinology*, remains an extremely complicated aspect of biology in the mid-twentieth century, but an extremely productive one as well.

Serology

The hormone-carrying function of blood was only one of the new virtues of the fluid discovered as the nineteenth century drew to its close. It served as a carrier of antibodies as well, and could thus serve as the general enemy of infection. (It is hard to believe now that a century and a half ago, physicians actually thought that the best way to help a sick patient was to deprive him of some of his blood.)

The use of blood against microorganisms came into its own with the work of two of the assistants of Koch (see page 103). These were the German bacteriologists, Emil Adolf von Behring (1854–1917) and Paul Ehrlich (1854–

1915). Von Behring discovered that it was possible to inject an animal with a particular germ and induce him to form antibodies against it which would be located in the liquid part of the blood ("blood serum"). If a sample of the blood were then taken from the animal, the serum containing the antibody could be injected into another animal, who would then be immune to the disease for a while at least.

It occurred to Von Behring to try this idea on the disease, diphtheria, which attacked children in particular and was almost sure death. If a child survived the disease it was immune thereafter, but why wait for the child to build its own antibodies in a race against the bacterial toxin? Why not prepare the antibodies in an animal first and then inject the antibody serum into the sick child. This was tried during a diphtheria epidemic in 1892 and the treatment was a success.

Ehrlich worked with Von Behring in this experiment and it was probably Ehrlich who worked out the actual dosages and techniques of treatment. The two men quarreled and Ehrlich worked independently thereafter, sharpening the methods of serum utilization to the point where he might be considered the real founder of *serology*, the study of techniques making use of blood serum. (Where these techniques involve the establishment of an immunity to a disease, the study may be called *immunology*.)

The Belgian bacteriologist, Jules Jean Baptiste Vincent Bordet (1870–1961), was another important serologist in the early days of that science. In 1898, while working in Paris under Mechnikov (see page 102), he discovered that if blood serum is heated to 55° C., the antibodies within it remain essentially unaffected, for they will still combine with certain chemicals ("antigens") with which they will also combine before heating. However, the ability of the serum to destroy bacteria is gone. Presumably some very fragile component, or group of components, of the serum must act as a complement for the antibody before the latter can react with bacteria. Bordet called this component "alexin," but Ehrlich named it, straightforwardly, "complement" and it is so known today.

In 1901, Bordet showed that when an antibody reacts with an antigen, complement is used up. This process of "complement fixation" proved important as a diagnostic device for syphilis. This was worked out in 1906 by the German bacteriologist, August von Wassermann (1866–1925), and is still known as the "Wassermann test."

In the Wassermann test, a patient's blood serum is allowed to react with certain antigens. If the antibody to the syphilis microorganism is present in the serum, the reaction takes place and complement is used up. The loss of complement is therefore indicative of syphilis. If complement is not lost, the reaction has not taken place, and syphilis is absent.

Blood Groups

The opening of the twentieth century saw a serological victory of a rather unexpected type. It dealt not with disease but with individual differences in human blood.

Physicians throughout history had occasionally tried to make up the blood loss in extensive hemorrhage by transferring blood from a healthy man, or even from an animal, into the veins of a patient. Despite occasional success, death was often hastened by such treatment, and most European nations had, by the end of the nineteenth century, prohibited attempts at such blood transfusions.

The Austrian physician, Karl Landsteiner (1868–1943), found the key to the problem. He discovered, in 1900, that human blood differed in the capacity of serum to agglutinate red blood corpuscles (that is, to cause them to clump together). One sample of blood serum might clump red blood corpuscles from person A but not from person B. Another sample of serum might, in reverse, clump the corpuscles from person B but not from person

A. Still another sample might clump both, and yet another might clump neither. By 1902, Landsteiner had clearly divided human blood into four "blood groups" or "blood types" which he named A, B, AB, and O.

Once this was done, it was a simple task to show that in certain combinations, transfusion was safe; while in others, the incoming red cells would be agglutinated, with possibly fatal results. Blood transfusion, based on a careful foreknowledge of blood groups of both patient and donor, became an important adjunct to medical practice at once.

Over the next forty years, Landsteiner and others discovered additional blood groups which did not affect transfusion. However, all these blood groups were inherited according to the Mendelian laws of inheritance (as was first shown in 1910) and they now form the basis for "paternity tests." Thus, two parents both of blood type A cannot have a child of blood type B, and such a child has either been switched in the hospital or has a father other than the suspected one.

Blood groups have come to offer a reasonable solution, too, for the age-old problem of "race." Men have always divided other men into groups, usually on some subjective and emotional basis that left their own group "superior." Even now, the layman tends to divide humanity into races on the basis of skin color.

The manner in which differences among individual human beings are gradual and not sharp, a matter of degree rather than of kind, was first made clear by a Belgian astronomer, Lambert Adolphe Jacques Quetelet (1796–1874). He applied statistical methods to the study of human beings and may therefore be considered a founder of *anthropology* (the study of the natural history of man).

He recorded the chest measurements of Scottish soldiers, the height of French Army draftees, and other such items and, by 1835, found that these varied from the average in the same manner that one would expect of the fall of dice or of the scatter of bullet holes about a bull's eye. In this way, randomness invaded the human realm and, in one more way, life was shown to follow the same laws that governed the inanimate universe.

A Swedish anatomist, Anders Adolf Retzius (1796–1860), tried to bend such anthropological measurements to the problem of race. The ratio of skull width to skull length, multiplied by 100, he called the "cranial index." A cranial index of less than 80 was "dolichocephalic" (long head) while one of over 80 was "brachycephalic" (wide head). In this way, Europeans could be divided into "Nordics" (tall and dolichocephalic); "Mediterraneans" (short and dolichocephalic), and "Alpines" (short and brachycephalic).

This is not really as satisfactory as it seems, for the differences are small, they do not apply well outside Europe and, finally, the cranial index is not really fixed and inborn but can be altered by vitamin deficiencies and by the environment to which the infant is subjected.

Once blood groups were discovered, however, the possibility of using these for classification proved attractive. For one thing, they are not a visible characteristic and therefore can't be used as a handy index for racism. They are truly inborn and are not affected by environment, and they are mixed freely down the generations since men and women are not influenced in the choice of mates by any consideration of blood groups (as they might be by visible characteristics).

No one blood group can be used to distinguish one race from another, but the average distributions of all the blood groups become significant when large numbers are compared. A leader in this branch of anthropology is the American immunologist, William Clouser Boyd (1903–   ). During the 1930s, he and his wife traveled to various parts of the earth, blood-typing the populations. From the data so obtained and from similar data obtained from others, Boyd, in 1956, was able to divide the human species into thirteen groups. Most of these followed logical geographic divisions. A surprise, however, was the existence of an "Early European" race character- ized by the presence of unusually high frequencies of a blood group termed "Rh minus." The Early Europeans were largely displaced by modern Europeans but a rem- nant (the Basques) persist even yet in the mountain fastnesses of the western Pyrenees.

Blood group frequencies can also be used to trace the course of prehistoric migrations, or even some that are not prehistoric. For instance, the percentage of blood type B is highest among the inhabitants of central Asia and falls off as one progresses westward and eastward. That it occurs at all in western Europe is thought by some to be the result of the periodic invasions of Europe during ancient and medieval times by central Asian nomads such as the Huns and Mongols.

Virus Diseases

But twentieth-century serology reserved its most spec- tacular successes for the battle with microorganisms of a type unknown to Pasteur and Koch in their day. Pasteur had failed to find the infective agent of rabies, a clearly infectious disease undoubtedly caused, according to his germ theory, by a microorganism. Pasteur suggested that the microorganism existed but that it was too small to be detected by the techniques of the time. In this, he turned out to be correct.

The fact that an infectious agent might be much smaller than ordinary bacteria was shown to be true in connection with a disease affecting the tobacco plant ("tobacco mosaic disease"). It was known that juice from diseased plants would infect healthy ones and, in 1892, the Russian botanist, Dmitri Iosifovich Ivanovski (1864– 1920), showed that the juice remained infective even after it had been passed through filters fine enough to keep any known bacterium from passing through. In 1895, this was discovered independently, by the Dutch botanist, Martinus Willem Beijerinck (1851–1931). Beijerinck named the infective agent a "filtrable virus" where virus simply means "poison." This marked the beginning of the science of *virology*.

Other diseases were proved to be caused by such filtrable viruses. The German bacteriologist, Friedrich August Johannes Löffler (1852–1915), was able to demonstrate, in 1898, that hoof-and-mouth disease was caused by a virus; and in 1901, Reed (see page 105) did the same for yellow fever. These were the first animal diseases shown to be virus-induced. Other diseases shown to be caused by viruses include poliomyelitis, typhus, measles, mumps, chicken pox, influenza, and the common cold.

A fine case, in this connection, of the biter bit, arose in 1915, when an English bacteriologist, Frederick William Twort (1877–1950), found that some of his bacterial colonies were turning foggy and then dissolving. He filtered these disappearing colonies and found that the filtrate contained something that caused normal colonies to dissolve. Apparently, bacteria themselves could suffer a virus disease and parasites were thus victimized by smaller parasites still. The Canadian bacteriologist, Félix Hubert d'Hérelle (1873–1949), made a similar discovery independently in 1917 and he named the bacteria-infesting viruses "bacteriophages" (bacteria-eaters).

In any listing of virus-caused diseases, cancer must remain a puzzle. Cancer has grown continually more important as a killer over the last century, for as other diseases are conquered, those that remain (cancer among them) claim a larger share of humanity for their own. The slowly inexorable advance of cancerous growths, the often lingering and painful death, have made cancer one of the prime terrors of mankind now.

During the initial successes of the germ theory, it had been thought that cancer might prove to be a bacterial disease, but no bacterium was found. After the existence of viruses was established, a cancer virus was sought for and not found either. This, combined with the fact that cancer was not infectious, caused many to believe that it was not a germ disease at all.

Although this may be so, it also remains true that although no general virus for the general disease has been discovered, particular viruslike agents have been discovered for particular types of cancer. In 1911, an American physician, Francis Peyton Rous (1879–   ), was studying a chicken with a kind of tumor called a "sarcoma." Among other things, he decided to test the sarcoma for virus content. He mashed it up and passed it through a filter. The filtrate, he found, would produce tumors in other chickens. He did not himself quite have the courage to call this the discovery of a virus, but others did.

For about a quarter of a century, the "Rous chicken sarcoma virus" was the only clear-cut example of anything like an infectious agent capable of inducing a cancer. In the 1930s and thereafter, however, further examples were discovered. Nevertheless, the matter remains unclear and the study of cancer (oncology) is still a major and frustrating branch of medical science.

While the physical nature of viruses remained unknown for some forty years after their discovery, this did not prevent logical steps being taken to treat virus diseases. In fact, smallpox, the first disease to be conquered by medical science, is a virus disease. Vaccination against smallpox encourages the body to form antibodies which will deal specifically with the smallpox virus and it is thus a kind of serological technique. Presumably, every virus disease could be countered by some serological treatment.

The difficulty, here, is that a strain of virus must be found which will produce no important symptoms and yet will spark the production of the necessary antibodies against the virulent strains (simulating the service performed by cowpox where smallpox is concerned). This sort of attack had been used by Pasteur in countering bacterial disease, but bacteria can be cultured without much trouble and can be easily treated in ways that will encourage the production of attenuated strains.

A virus, unfortunately, can live only in living cells and this increases the difficulty of the problem. Thus, the South African microbiologist, Max Theiler (1899–), produced a vaccine against yellow fever in the 1930s only after he had painstakingly transferred the yellow-fever virus first to monkeys and then to mice. In mice, it developed as an encephalitis, or brain inflammation. He passed the virus from mouse to mouse and then, eventually, back to monkeys. By then, it was an attenuated virus, producing only the feeblest yellow-fever attack, but inducing full immunity to the most virulent strains of the virus.

Meanwhile, though, a living analog of Koch's nutrient broths was discovered by the American physician, Ernest William Goodpasture (1886–1960). In 1931, he introduced the use of living chick embryos as a nutrient for viruses. If the top of the shell is removed, the rest of the shell serves as a natural Petri dish (see page 104). By 1937, a still safer yellow-fever vaccine was produced by Theiler after he had selected a nonvirulent virus strain from among those he had passed along from chick embryo to chick embryo in nearly two hundred transplants.

The most spectacular accomplishment of the new serological techniques was in connection with poliomyelitis. The virus was first isolated in 1908 by Landsteiner (see page 129) who was also the first to transmit the disease to monkeys. Monkeys are expensive and are difficult experimental animals, however, and to find a nonvirulent strain by infecting crowds of monkeys is impractical.

The American microbiologist, John Franklin Enders (1897–), with two young associates, Thomas Huckle Weller (1915–) and Frederick Chapman Robbins (1916–), attempted, in 1948, to culture virus in mashed-up chick embryos, bathed in blood. Attempts of this sort had been made earlier by others but always the effort had failed, since whether the virus multiplied or not, the culture was drowned out by the rapidly multiplying bacteria. Enders, however, had the notion of adding the recently developed penicillin to his cultures. This stopped bacterial growth without affecting the virus, and in this way he managed to culture the mumps virus successfully.

He next tried this technique on the poliomyelitis virus and, in 1949, succeeded again. Now it was possible to culture the virus easily and in quantity so that one might search among hundreds of strains for an attenuated one of the proper characteristics. The Polish-American microbiologist, Albert Bruce Sabin (1906–   ), had, by 1957, discovered an attenuated strain of poliomyelitis virus for each of the three varieties of the disease, and had produced successful vaccines against the disease.

In similar fashion, Enders and his associate Samuel Lawrence Katz (1927–   ) developed an attenuated strain of measles virus in the early 1960s, which may serve as a vaccine to end the threat of that children's disease.

Allergy

The body's mechanism of immunity is not always utilized in a manner which seems to us to be beneficial. The body can develop the ability to produce antibodies against any foreign protein, even some which might be thought to be harmless. Once the body is "sensitized" in this fashion, it will react to contact with the protein in various distressing ways—swollen mucus membranes in the nose, overproduction of mucus, coughing, sneezing, watering of the eyes, contraction of the bronchioles in the lungs ("asthma"). In general, the body has an "allergy." Quite commonly, the allergy is to some food component, so that the sufferer will break out in itchy blotches ("hives") if he is not careful with his diet, or he will react to plant pollen and will suffer from the misnamed "hay fever" at certain times of the year.

Since antibodies will be formed against the proteins of other human beings (even these are sufficiently alien), it follows that each human being (multiple births excepted) is a chemical individual. It is not practical, for that reason, to try to graft skin, or some organ, from one person to another. Even where infection is prevented by modern techniques, the patient receiving the graft develops the antibodies necessary to fight it off. This is analogous to the difficulties of transfusion, but with the problems much intensified, for human tissues cannot be classified into a few broad types as human blood can.

This is unfortunate, for biologists have learned to keep portions of the body alive for periods of time. A heart removed from an experimental animal can be kept beating for a while without much trouble and, in 1880, the English physician, Sydney Ringer (1834–1910), developed a solution containing various inorganic salts in the proportions usually found in blood. This would act as an artificial circulating fluid to keep an isolated organ alive for quite respectably long periods.

The art of keeping organs alive by means of nutrient solutions of the proper ionic content was developed to a fine art by the French-American surgeon, Alexis Carrel (1873–1944). He kept a piece of embryonic chicken heart alive and growing (it had to be periodically trimmed) for over twenty years.

It follows then that the possibility of organ transplantation, when such an organ is required to save a life, would be bright, were it not for the adverse antibody response. Even so, some transplantations, such as the cornea of the eye, can be made routinely, while, in the 1960s, successful kidney transplants have occasionally been managed.

In 1949, the Australian physician, Frank Macfarlane Burnet (1899–), suggested that the ability of an or-

ganism to form antibodies against foreign proteins might
not be inborn after all, but might develop only in the
course of life; though perhaps very early in life. An English biologist, Peter Brian Medawar (1915–   ), tested the
suggestion by inoculating the embryos of mice with tissue
cells from mice of another strain (without recent common ancestors). If the embryos had not yet gained the
ability to form antibodies, then by the time they reached
independent life and could form them, it might be that
the particular foreign proteins with which they had been
inoculated would no longer seem foreign. This turned out,
indeed, to be the case, and in adult life, the mice, having
been inoculated in embryo were able to accept skin grafts
from a strain where, without inoculation, they would not
have been able to do so.

In 1961, it was discovered that the thymus gland, hitherto not known to have any function, was the source of
the body's ability to form antibodies. The thymus produces lymphocytes (a variety of white blood corpuscle)
whose function it is to form antibodies. Shortly after
birth, the lymphocytes produced by the thymus travel
to the lymph nodes and into the blood stream. After a
while, the lymph nodes can continue on their own and
at puberty, the thymus gland, its job done, shrivels and
shrinks to nothing. The effect of this new discovery on
possible organ transplantation remains to be seen.

CHAPTER 12

Metabolism

Chemotherapy

The drive against the bacterial diseases is in some ways simpler than that against the virus diseases. As explained in the previous chapter, bacteria are easier to culture. In addition, bacteria are more vulnerable. Living as they do outside cells, they are capable of doing damage by successfully competing for food or by liberating toxic substances. However, their chemical machinery, or metabolism, is generally different from that of the cells of the host in at least some respects. There is always the chance, therefore, that they might be vulnerable to chemicals that would disorder their metabolism without seriously affecting the metabolism of the host cells.

The use of chemical remedies against disease dates back into prehistory. Down to modern times, the "herbwoman" and her concoctions, handed down empirically over the generations, have on occasion done some good. The use of quinine against the malaria parasite is the best-known example of a chemical that began as a folk remedy and was later accepted by the medical profession.

With the coming of synthetic organic chemicals that did not occur in nature, however, the possibility arose that many more such specifics might be found; that every disease might have its particular chemical remedy. The great early protagonist of this view was Ehrlich (see pp. 127–28), who spoke of such chemical remedies as "magic bullets" that sought out the germ and slew it while leaving the body cells in peace.

He had worked with dyes that stained bacteria and, since these dyes specifically combined with some constituent of the bacterial cell, they ought to damage the bacterial cell's working mechanism. He hoped to find one that would do this without harming ordinary cells too badly. Indeed, he did discover a dye, "trypan red," which helped destroy the trypanosomes (a protozoan, rather than a bacterium, but the principle was the same) that caused such diseases as sleeping sickness.

Ehrlich kept looking for something better. He decided that the action of trypan red was caused by the nitrogen-atom combinations it contained. Arsenic atoms resemble nitrogen atoms in chemical properties but, in general, introduce a more poisonous quality into compounds. Ehrlich was led by that into a consideration of arsenicals. He began to try all the arsenic-containing organic compounds he could find or synthesize, hundreds of them, one after the other.

In 1909, one of his assistants discovered that the compound numbered 606, tried against the trypanosome and found wanting, was very effective on the causative agent of syphilis. Ehrlich named the chemical "Salvarsan" (though a more frequently used synonym, nowadays, is "arsphenamine") and spent the remainder of his life improving the technique for using it to cure syphilis.

Trypan red and Salvarsan marked the beginning of modern *chemotherapy* (the treatment of disease by chemicals, a word coined by Ehrlich) and for a while hopes were high that other diseases would be treated in similar fashion. Unfortunately, for twenty-five years after the discovery of arsphenamine's effect, the vast list of synthetic organic chemicals seemed to offer nothing more.

But then came another stroke of good fortune. A German biochemist and physician, Gerhard Domagk (1895–), working for a dye firm, began a systematic survey of new dyes with a view to finding possible medical applications for some of them. One of the dyes was a newly synthesized orange-red compound with the trademark "Prontosil." In 1932, Domagk found that injections of the dye had a powerful effect on streptococcus infections in mice.

He quickly had a chance to try it on humans. His young daughter had been infected by streptococci following the prick of a needle. No treatment did any good until Domagk in desperation injected large quantities of Prontosil. She recovered dramatically and, by 1935, the world had learned of the new drug.

It was not long before it was recognized by a group of French bacteriologists, that not all of the molecule of Prontosil was needed for the antibacterial effect to be evident. A mere portion of it, called "sulfanilamide" (a compound known to chemists since 1908) was the effective principle. The use of sulfanilamide and related "sulfa" compounds inaugurated the era of the "wonder drugs." A number of infectious diseases, notably some varieties of pneumonia, suddenly lost their terrors.

Antibiotics and Pesticides

And yet the greatest successes of chemotherapy were not to lie with synthetic compounds like arsphenamine and sulfanilamide but with natural products. A French-American microbiologist, René Jules Dubos (1901–   ), was interested in soil microorganisms. After all, the soil received the dead bodies of animals with every conceivable disease and, except in rare cases, it was not itself a reservoir of infection. Apparently, there were agents within the soil that were antibacterial. (Such agents later came to be called "antibiotics" meaning "against life.")

In 1939, Dubos isolated the first of the antibiotics, "tyrothricin," from a soil bacterium. It was not a very effective antibiotic, but it revived interest in an observation made by a Scottish bacteriologist, Alexander Fleming (1881–1955), over a decade earlier.

In 1928, Fleming had left a culture of staphylococcus germs uncovered for some days. He was through with it and was about to discard the dish containing the culture when he noticed that some specks of mold had fallen into it and that around every speck, the bacterial colony had dissolved away for a short distance.

Fleming isolated the mold and eventually identified it as one called *Penicillium notatum*, a mold closely related to the common variety often found growing on stale bread. Fleming decided that the mold liberated some compound which, at the very least, inhibited bacterial growth. He called the substance, whatever it might be, "penicillin." He investigated it to the point of showing that it would affect some bacteria and not others and that it was not harmful to white blood corpuscles and, therefore, possibly not harmful to other human cells. Here he had to let his efforts stop.

However, 1939 saw interest in antibiotics (of which penicillin was clearly an example) bound upward, thanks to Dubos' work. In addition, the coming of World War II meant that any weapon to combat infected wounds would be welcome. An Australian-English pathologist, Howard Walter Florey (1898–), together with a German-English biochemist, Ernst Boris Chain (1906–), tackled the problem of isolating penicillin, determining its structure and learning how to produce it in quantity. By war's end, they headed a large Anglo-American research team and succeeded brilliantly. Penicillin became and even yet remains the work horse of the doctor's weapon against infection.

After the war, other antibiotics were sought for and found. The Russian-American bacteriologist, Selman Abraham Waksman (1888–), went through soil microorganisms as systematically as Ehrlich had gone through synthetics. In 1943, he isolated an antibiotic that was effective against many bacteria that were unaffected by penicillin. In 1945 it went on the market as "streptomycin." (It was Waksman, by the way, who coined the word "antibiotic.")

In the early 1950s, the "broad-spectrum antibiotics" (those affecting a particularly wide range of bacteria) were discovered. These are the "tetracyclines," best known to the public by such trade-marks as "Achromycin" and "Aureomycin."

Bacterial diseases have been brought under control, as a result of the discovery of antibiotics, to a degree that would have seemed overoptimistic only a generation ago. Nevertheless, the future is not entirely rosy. Natural selection marks for survival those strains of bacteria that have a natural resistance to antibiotics. Therefore, with time, particular antibiotics become less effective. New antibiotics will certainly be discovered so that all will not be lost. Nevertheless, all will not be won either, and may never be.

The various chemotherapeutic agents do not, in general, affect viruses. These multiply inside living cells and can be killed by chemical attack only if the cell itself is killed. A more indirect attack, however, may be successful, for a chemical may kill not the virus itself but the multi-cellular creature that carries the virus.

The virus of typhus fever is carried by the body louse, for instance, a creature much harder to get rid of (since it is so closely bound to the unwashed, old-clothed human body) than is the free-living mosquito. Yellow fever and malaria can be handled by mosquito-control but typhus fever remained mightily dangerous and in Russia and the Balkans during World War I, it was more deadly to both sides, on occasion, than the enemy artillery was.

In 1935, however, a Swiss chemist, Paul Müller (1899–), began a research program designed to discover some organic compound that would kill insects quickly without seriously affecting other animal life. In September 1939, he found that "dichlorodiphenyltrichloroethane" (usually abbreviated as "DDT"), first synthesized in 1873, would do the trick.

In 1942, it began to be produced commercially and, in 1943, it was used during a typhus epidemic that broke out in Naples soon after it had been captured by Anglo-American forces. The population was sprayed with DDT, the body lice died, and for the first time in history, a winter epidemic of typhus was stopped in its tracks. A similar epidemic was stopped in Japan in late 1945, after American forces had occupied the nation.

Since World War II, DDT and other organic insecticides have been used against insects not only to prevent disease but to keep down the havoc they wreak against man's food crops. Weed killers have also been devised and these may be lumped with insect killers under the heading of "pesticides."

Here again, insects develop resistant strains and particular pesticides become less effective with time. In addition, many fear that the indiscriminate use of pesticides needlessly kills many forms of life that are not harmful to man, and upsets the balance of nature in a way that will, in the end, do far more harm than good.

This is a serious problem. The study of the interrelationships of life forms ("ecology") is a difficult and intricate one and much remains to be understood here. Mankind is continually altering the environment in ways that are intended for short-term benefit, but we can never be entirely sure that the distortions introduced into the web of life, even when seemingly unimportant, may not be to our long-term harm.

Metabolic Intermediates

The effect of chemotherapeutic agents on insects, weeds, and microorganisms is that of interfering with the pattern of metabolism—sabotage of the organisms' chemical machinery, in other words. The search for such agents is increasingly rationalized by growing knowledge concerning the details of metabolism.

In this respect, the English biochemist, Arthur Harden (1865–1940), led the way. He was interested in the enzymes in yeast extract (the extract which Buchner had shown to be as efficient at breaking down sugar as the yeast cells themselves—see page 96). In 1905, Harden noted that a sample of extract broke down sugar and produced carbon dioxide quite rapidly at first, but that with time, the rate of activity dropped off. This might seem to be due to the gradual wearing out of the enzymes in the extract, but Harden showed this was not the case. If he added small quantities of sodium phosphate (a simple inorganic compound) to the solution, the enzyme went back to work as hard as ever.

Since the concentration of the inorganic phosphate decreased as the enzyme reaction proceeded, Harden searched for some organic phosphate formed from it and located that in the form of a sugar molecule to which two phosphate groups had become attached. This was the beginning of the study of "intermediary metabolism"; the search for the numerous compounds formed as intermediates (sometimes very briefly lived ones) in the course of the chemical reactions going on in living tissue.

Some of the main lines of this search can be sketched out. The German biochemist, Otto Fritz Meyerhof (1884–1951), in 1918 and the years thereafter, showed that in muscle contraction, glycogen (a form of starch) disappeared, while lactic acid appeared in corresponding amounts. In the process, oxygen was not consumed, so that energy was obtained without oxygen. Then, when the muscle rested after work, some of the lactic acid was oxidized (molecular oxygen being *then* consumed to pay off an "oxygen debt"). The energy so developed made it possible for the major portion of the lactic acid to be reconverted to glycogen. The English physiologist, Archibald Vivian Hill (1886–   ), came to the same conclusions at about the same time, by making delicate measurements of the heat developed by contracting muscle.

The details of this conversion of glycogen to lactic acid were worked out during the 1930s by the Czech-American biochemists Carl Ferdinand Cori (1896–) and his wife, Gerty Theresa Cori (1896–1957). They isolated a hitherto unknown compound from muscle tissue, glucose-1-phosphate (still called "Cori ester") and showed that it was the first product of glycogen breakdown. Painstakingly, they followed glucose-1-phosphate through a series of other changes and fitted each intermediate into the breakdown chain. One of the intermediates proved to be the sugar phosphate first detected by Harden a generation earlier.

The fact that Harden and the Coris came across phosphate containing organic compounds in their search for intermediates was significant. Throughout the first third of the twentieth century, the phosphate group was found to play an important part in one biochemical mechanism after another. The German-American biochemist, Fritz Albert Lipmann (1899–), explained this by showing that phosphate groups could occur within molecules in one of two types of arrangement: low energy and high energy. When molecules of starch or fat were broken down, the energy liberated was used to convert low-energy phosphates to high-energy phosphates. In this way, the energy was stored in convenient chemical form. The breakdown of one high-energy phosphate liberated just enough energy to bring about the various energy-consuming chemical changes in the body.

Meanwhile, those steps in the breakdown of glycogen that lay beyond lactic acid and that did require oxygen could be studied by means of a new technique developed by a German biochemist, Otto Heinrich Warburg (1883–). In 1923, he devised a method for preparing thin slices of tissue (still alive and absorbing oxygen) and measuring their oxygen uptake. He used a small flask attached to a thin U-shaped tube. In the bottom of the tube was a colored solution. Carbon dioxide produced by the tissue was absorbed by a small well of alkaline solution within the flask. As oxygen was absorbed without being replaced in the air by carbon dioxide, a partial vacuum was produced in the flask and the liquid in the U-tube was sucked upward toward the flask. The rate of level change of the fluid, measured under carefully controlled conditions, yielded the rate of oxygen uptake.

The influence of different compounds on this rate of uptake could then be studied. If a particular compound restored the rate after it had fallen off, it might be taken to be an intermediate in the series of reactions involved in oxygen uptake. The Hungarian biochemist, Albert

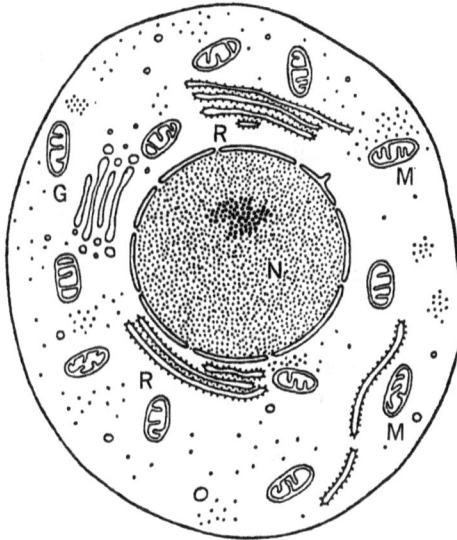

FIGURE 5. The generalized structure of a cell, seen through an electron microscope. N is the nucleus, the darker area within that the nucleolus; M denotes the mitochondria, G the Golgi bodies, and R the reticulum of particle-covered membranes. The dots here and elsewhere in the cell represent centers of protein synthesis and are known as ribosomes.

Szent-Györgyi (1893–) and the German-British bio-chemist, Hans Adolf Krebs (1900–), were active in this respect. Krebs had, indeed, by 1940, worked out all the main steps in the conversion of lactic acid to carbon dioxide and water, and this sequence of reactions is often called the "Krebs cycle." Earlier, during the 1930s, Krebs had also worked out the main steps in the formation of the waste product, urea, from the amino acid building blocks of proteins. This removed the nitrogen and the remainder of the amino acid molecules could, as Rubner had shown almost a half-century earlier (see page 89), be broken down to yield energy.

Hand in hand with this increase of knowledge concerning the internal chemistry of the cell came an increase of knowledge concerning the fine structure of the cell. New techniques for the purpose were developed. In the early 1930s, the first "electron microscope" was built. This magnified by focusing electron beams rather than light waves and the result was far greater magnification than was possible with ordinary microscopes. The Russian-American physicist, Vladimir Kosma Zworykin (1889–), modified and refined the instrument to the point where it became a practical and useful tool in cytology. Particles no larger than very large molecules could be made out and the protoplasm of the cell was found to be an almost bewildering complex of small but highly organized structures called "organelles" or "particulates."

Techniques were devised, in the 1940s, whereby cells would be minced up and the various organelles separated according to size. Among the larger and more easily studied of these are the "mitochondria" (singular, "mitochondrion"). A typical liver cell will contain about a thousand mitochondria, each a rodlike object, about two to five thousandths of a millimeter long. These were investigated in particular detail by the American biochemist, David Ezra Green (1910–), and his associates and were found by them to be the site of the reactions of the

Krebs cycle. Indeed, all the reactions involving the use of molecular oxygen took place there, with the enzymes catalyzing the various reactions arranged in appropriate organization within each mitochondrion. The little organelle thus proved to be "the powerhouse of the cell."

Radioactive Isotopes

The manner in which the intricate chain of metabolic reactions could be worked out was greatly facilitated by the use of special varieties of atoms called "isotopes." During the first third of the twentieth century, physicists had discovered that most elements consisted of several such varieties. The body did not distinguish among them to any great degree but laboratory apparatus had been devised which could do so.

The German-American biochemist, Rudolf Schoenheimer (1898–1941), was the first to make large-scale use of isotopes in biochemical research. By 1935, a rare isotope of hydrogen, hydrogen-2, twice as heavy as ordinary hydrogen, was available in reasonable quantities. Schoenheimer used it to synthesize fat molecules that contained the rare hydrogen-2 ("heavy hydrogen" or "deuterium") in place of the ordinary hydrogen-1. These were incorporated into the diet of laboratory animals, whose tissues treated the heavy-hydrogen fat much as they would ordinary fat. Analysis of the body fat of the animals for hydrogen-2 content threw new and startling light on metabolism.

It was believed at the time that the fat stores of an organism were generally immobile, and were only mobilized in time of famine. However, when Schoenheimer fed rats on his hydrogen-2 fat, then analyzed the fat stores, he found that at the end of four days, the tissue fat contained nearly half the hydrogen-2 that had been fed the animal. In other words, ingested fat was stored and stored fat was used. There was a rapid turnover and the body constituents were undergoing constant change.

Schoenheimer went on to use nitrogen-15 ("heavy nitrogen") to tag amino acids. He would feed rats on a mixture of amino acids, only one of which might be tagged, and then find that after a short time, all the different amino acids in the rat were tagged. Here, too, there was constant action. Molecules were rapidly changing and shifting even though the over-all movement might be small.

In principle, one might follow the exact sequence of changes by detecting the various compounds in which the isotope appeared, one after the other. This was most easily done with radioactive isotopes, atom varieties which were unusual not only in weight but in the fact that they broke down, liberating fast-moving energetic particles. These particles were easily detected so that very small quantities of radioactive isotopes would suffice for experimentation. After World War II, radioactive isotopes were prepared in quantity by means of nuclear reactors. In addition, a radioactive isotope of carbon ("carbon-14") was discovered and found to be particularly useful.

Radioactive isotopes, for instance, enabled the American biochemist, Melvin Calvin (1911–), to work out many of the fine details of the sequence of reactions involved in photosynthesis; that is, the manner in which green plants converted sunlight into chemical energy and supplied the animal world with food and oxygen.

Calvin allowed microscopic plant cells access to carbon dioxide in the light for only a few seconds, then killed the cells. Presumably only the first stages of the photosynthetic reaction chain would have an opportunity to be completed. The cells were mashed up and separated into their components by a technique called paper chromatography which will be described in the next chapter. Which of these components, however, represented the first stage of photosynthesis and which were present for other reasons?

Calvin could tell because the carbon dioxide to which the plant cells had had access contained radioactive carbon-14 in its molecules. Any substance formed from that carbon dioxide by photosynthesis would itself be radioactive and would be easily detected. This was the starting point of a series of researches through the 1950s that produced a useful scheme of the main steps in photosynthesis.

CHAPTER 13

Molecular Biology: Protein

Enzymes and Coenzymes

The pattern of metabolism, sketched out in finer and finer detail as the mid-twentieth century passed, was, in a way, an expression of the enzymatic makeup of the cell. Each metabolic reaction is catalyzed by a particular enzyme and the nature of the pattern is determined by the nature and concentration of the enzymes present. To understand metabolism, therefore, it was desirable to understand enzymes.

Harden, who had begun the twentieth-century unravelment of intermediary metabolism (see page 145), also unfolded a new aspect of enzymes. In 1904, he placed an extract of yeast inside a bag made of a semipermeable membrane (one through which small molecules might pass but not large ones) and placed it in water. The small molecules in the extract passed through and, after a while, the yeast extract could no longer break down sugar.

This could not be because the enzyme itself had passed through, since the water outside the bag could not break down sugar either. However, if the water outside were added to the extract inside, the mixture could break down sugar. The conclusion was that an enzyme (itself a large molecule unable to pass through a membrane) might yet include a relatively small molecule, loosely bound and therefore capable of breaking free and passing through the membrane, as part of its structure and essential to its function. The small, loosely bound portion came to be called a "coenzyme."

The structure of Harden's coenzyme was worked out, during the 1920s, by the German-Swedish chemist, Hans Karl von Euler-Chelpin (1873–   ). Other enzymes were found to include coenzyme portions and the structure of a number of these was elucidated during the 1930s. As the molecular structure of vitamins was also determined in that decade, it became quite apparent that many of the coenzymes contained vitaminlike structures as part of their molecules.

Apparently, then, vitamins represented those portions of coenzymes which the body could not manufacture for itself and which, therefore, had to be present, intact, in the diet. Without the vitamins, the coenzymes could not be formed; without the coenzymes, certain enzymes were ineffective and the metabolic pattern was badly upset. The result was a vitamin-deficiency disease and, eventually, death.

Since enzymes are catalysts, needed by the body only in small quantities, coenzymes (and vitamins, too) are needed in small quantities only. This explains why a dietary component, present only in traces, may yet be essential to life. It was easy to see that minerals needed in traces, such as copper, cobalt, molybdenum, and zinc, must also form essential parts of an enzymatic structure, and enzymes containing one or more atoms of such elements have indeed been isolated.

But what of the enzyme itself? Throughout the nineteenth century, it had been a mysterious entity, visible only through its effects. The German-American chemist, Leonor Michaelis (1875–1949), brought it down to earth in a way by treating it according to physical-chemical principles. He applied the rules of chemical kinetics (a branch of physical chemistry that deals with the rates of reactions) and, in 1913, was able to derive an equation that described the manner in which the rate of an enzyme-catalyzed reaction varies under certain set circumstances. To work out this equation, he postulated an intermediate combination of the enzyme and the substance whose reaction it catalyzed. This sort of treatment emphasized that enzymes were molecules that obeyed the physical-chemical laws to which other molecules were subject.

But what kind of a molecule was it? To be sure, it was strongly suspected of being a protein, for an enzyme solution easily lost its activity through gentle heating and only protein molecules were known to be so fragile. This, however, was only supposed and not proven, and during the 1920s, the German chemist, Richard Willstätter (1872–1942), advanced reasons for believing that enzymes were *not* proteins. His reasoning, as it turned out, was fallacious, but his prestige was great enough to lend his opinion considerable weight.

In 1926, however, the possibility that enzymes were proteins was raised again by an American biochemist, James Batchellor Sumner (1887–1955). In that year, Sumner was extracting the enzyme content of jack beans, the enzyme involved being "urease," one which catalyzed the breakdown of urea to ammonia and carbon dioxide.

In performing his extraction, Sumner found that at one point he obtained a number of tiny crystals. He isolated the crystals, dissolved them, and found he had a solution with concentrated urease activity. Try as he might, he could not separate the enzyme activity from the crystals. The crystals *were* the enzyme and all his tests further agreed on the fact that the crystals were also protein. Urease, in short, was the first enzyme ever to be prepared in crystalline form, and the first enzyme to be shown, incontrovertibly, to be a protein.

It further confirmation was wanting, or if the rule was suspected to be not general, the work of the American biochemist, John Howard Northrop (1891–   ), finished matters. In 1930, he crystallized pepsin, the protein-splitting enzyme in gastric juice; in 1932, he crystallized trypsin and, in 1935, chymotrypsin, both protein-splitting enzymes from pancreatic juice. These proved to be protein, too. Since then, dozens of enzymes have been crystallized and all have proved to be proteins.

By the mid-1930s then, the problem of enzymes had clearly merged with the general problem of proteins.

Electrophoresis and X-ray Diffraction

The development of new chemical and physical tools during the first half of the twentieth century made it possible for biochemists to probe with increasing finesse the very large protein molecules that seemed to be the very essence of life. In fact, what amounted to a new field of science, one that combined physics, chemistry, and biology, took for its realm of study the analysis of the fine structure and detailed functioning of the giant molecules of life. This new field, *molecular biology*, has become particularly important (and, indeed, quite spectacular in its achievements) since World War II, and has tended to overshadow the remainder of biology.

In 1923, the Swedish chemist, Theodor Svedberg (1884–   ), introduced a powerful method for determining the size of protein molecules. This was an "ultra-centrifuge," a spinning vessel that produced a centrifugal force hundreds of thousands of times as intense as that of ordinary gravity. The thermal agitation of molecules of water at ordinary temperature suffice to keep the giant protein molecules in even suspension against the pull of ordinary gravity but not against such a centrifugal force. In the whirling ultracentrifuge, protein molecules begin to settle out, or "to sediment." From the sedimentation rate, the molecular weight of protein molecules can be determined. A protein of average size, such as hemoglobin, the red coloring matter of blood, has a molecular weight of 67,000. It is 3700 times as large as a water molecule, which has a molecular weight of only 18. Other protein molecules are larger still, with molecular weights in the hundreds of thousands.

The size and complexity of the protein molecule means that there is ample room on the molecular surface for atom groupings capable of carrying electric charges. Each protein has its own pattern of positive and negative charges on its molecular surface—a pattern different from that of any other protein and one capable of changing in fixed manner with changes in the acidity of the surrounding medium.

If a protein solution is placed in an electric field, the individual protein molecules travel toward either the positive or negative electrode at a fixed speed dictated by the pattern of the electric charge, the size and shape of the molecule and so on. No two varieties of protein would travel at precisely the same speed under all conditions.

In 1937, the Swedish chemist, Arne Wilhelm Kaurin Tiselius (1902–   ), a student of Svedberg's, devised an apparatus to take advantage of this. This consisted of a special tube arranged like a rectangular U, within which a protein mixture could move in response to an electric field. (Such motion is called "electrophoresis.") Since the various components of the mixture moved each at its own rate, there was a gradual separation. The rectangular-U tube consisted of portions that fitted together at specially ground joints, and these portions could be slid apart. Matters could be arranged so that one of the mixture of proteins would be present in one component of the chambers and could thus be separated from the rest.

Furthermore, by the use of appropriate cylindrical lenses, it became possible to follow the process of separation by taking advantage of changes in the way light was refracted on passing through the suspended mixture as the protein concentration changed. The changes in refraction could be photographed as a wavelike pattern which could then be used to calculate the quantity of each type of protein present in the mixture.

The proteins in blood plasma, in particular, were subjected to electrophoresis and studied. They were separated into numerous fractions, including an albumin, and three groups of globulins, distinguished by Greek letters as alpha, beta, and gamma. The gamma-globulin fraction was found to contain the antibodies. During the 1940s, methods were devised to produce the different protein fractions in quantity.

Ultracentrifugation and electrophoresis depended upon the properties of the protein molecule as a whole. The use of X rays enabled the biochemist to probe within the molecule. An X-ray beam is scattered in passing through matter, and where the constituent particles of matter are arranged in regular ranks and files (as atoms are arranged within crystals) the scattering is regular, too. An X-ray beam impinging upon a photographic film, after being scattered by a crystal, appears as a symmetrical pattern of dots from which the arrangement and distance of separation of the atoms within a crystal may be deduced.

It often happens that large molecules are built up of smaller units which are arranged regularly within the molecules. This is true, for instance, of proteins, which are built up of amino acids. The regular arrangement of amino acids within a protein molecule is reflected in the manner in which an X-ray beam is scattered. The resulting scattering is less clear cut than that produced by a crystal, but it is capable of analysis. In the early 1930s, the general spacing of amino acid units was deduced.

This was sharpened in 1951, when the American chemist, Linus Pauling (1901–), worked out the amino acid arrangement and showed that the chain of these units was arranged in the form of a helix. (A helix is the shape of what is usually called a spiral staircase.)

As men probed more and more deeply into the details of protein structure, it became necessary to deal with more and more complicated X-ray data, and the necessary mathematical computations grew long-winded and intractable, reaching a point where their detailed solution by the unaided human mind was impractical. Fortunately, by the 1950s, electronic computers had been developed which could perform routine computation of immense length in very little time.

The computer was first put to use in this manner in a problem involving not a protein, but a vitamin. In 1924, two American physicians, George Richards Minot (1885–1950) and William Parry Murphy (1892–), had discovered that the regular feeding of liver kept patients from dying of a disease called "pernicious anemia." The presence of a vitamin was suspected. It was named vitamin B_{12} and in 1948 it was finally isolated. It proved to have a very complicated molecule built up of 183 atoms of six different elements. With the new physical techniques and the aid of a computer, the detailed structure of the vitamin was worked out in 1956. Because it was found to contain a cyanide group, a cobalt atom, and an amine group (among numerous other structures), it was renamed "cyanocobalamine."

It was inevitable that computers be applied to the diffraction patterns set up by proteins. Using X-ray diffraction and computers, the Austrian-British biochemist, Max Ferdinand Perutz (1914–) and the English biochemist, John Cowdery Kendrew (1917–), were able to announce, in 1960, a complete three-dimensional picture of the molecule of myoglobin (a muscle protein something like hemoglobin but one quarter the size) with every amino acid in place.

Chromatography

The use of physical methods, such as X-ray diffraction, to work out the detailed structure of a large molecule, is immeasurably aided if chemists have already determined the chemical nature of the subunits of the molecule and have obtained a general notion of their arrangement. If this is done, the number of possibilities into which the esoteric diffraction data need be fitted, is cut down to a practical size.

In the case of proteins, chemical progress was slow at first. The men of the nineteenth century had only been able to show that the protein molecule was built up out of amino acids. As the twentieth century opened, the German chemist, Emil Hermann Fischer (1852–1919), demonstrated the manner in which amino acids were combined within the protein molecule. In 1907, he was even able to put together fifteen molecules of one amino acid and three of another to form a very simple eighteen-unit proteinlike substance.

But what was the exact structure of the far more complicated protein molecules occurring in nature? To begin with, what was the exact number of each type of amino acid present in a given protein molecule? The straightforward method of answering that question would have been to break up the protein molecule into a mixture of individual amino acids and then to determine the relative quantities of each component by the methods of chemical analysis.

This procedure was impractical, however, for the chemists of Emil Fischer's day. Some of the amino acids were sufficiently similar in structure to defeat ordinary chemical methods intended for use in differentiating among them.

The answer to the problem came through a technique, the ancestor of which first saw the light of day in 1906, thanks to the labors of a Russian botanist, Mikhail Semenovich Tsvett (1872–1919). He was working with plant pigments and found a complex mixture on his hands, one made up of compounds so similar as to be separable only with the greatest difficulty by ordinary chemical methods. It occurred to him, however, to let a solution of the mixture trickle down a tube of powdered alumina. The different substances in the pigment mixture held to the surface of the powder particles with different degrees of strength. As the mixture was washed downward with fresh solvent, they separated; those components of the mixture which held with less strength being washed down further; in the end, the mixture was separated into individual pigments each with its own shade of color. The fact of separation was "written in color" and Tsvett named the technique, from the Greek for that phrase, as "chromatography."

Tsvett's work roused little interest at the time, but in the 1920s Willstätter (see page 153) reintroduced it and made it popular. Chromatography came to have a wide and varied use in the separation of complex mixtures. In the form of a tube of powder, however, it could only with difficulty be applied to very small quantities of mixture. Something still more powerful was needed.

The necessary modification came in 1944 and revolutionized biochemical technique. In that year, the English biochemists, Archer John Porter Martin (1910–   ) and Richard Laurence Millington Synge (1914–   ), worked out a technique for carrying on chromatography on simple filter paper.

A drop of an amino acid mixture was allowed to dry near the bottom of a strip of filter paper and a particular solvent (into which the bottom edge of the strip could be dipped) was then allowed to creep up the strip by capillary action. As the creeping solvent passed the dried mixture, the individual amino acids contained therein crept up with the solvent, but each at its own characteristic rate. In the end, the amino acids were separated. Their position on the paper could be detected by some suitable physical or chemical method and matched against the position of individual amino acids treated separately in the same way on other pieces of paper. The quantity of amino acids in each spot could be determined without much difficulty.

This technique of "paper chromatography" proved an instant success. Simply and inexpensively, without elaborate equipment, it neatly separated tiny amounts of complex mixtures. The technique was quickly applied to virtually every branch of biochemistry—to Calvin's work on mixtures in photosynthesizing plant cells (see page 150), for instance—until research without the technique has become virtually unthinkable.

In particular, paper chromatography made it possible to determine the exact number of the different amino acids present in a particular protein. Protein after protein came to be characterized by the number of each of its constituent amino acids, as an ordinary compound might be identified by the number of atoms of each of its constituent elements.

Amino Acid Arrangement

This, however, was still not enough. After all, chemists are interested not only in the number of atoms in an ordinary compound, but in their arrangement as well; and so it is with the amino acids in protein molecules (see Figure 6). The question of arrangement is a difficult one, though. With even a few dozen amino acids in a molecule, the number of possible different arrangements is astronomical, and with 500-plus amino acids present (as in the molecule of hemoglobin, which is only of average size for a protein) the different arrangements possible must be represented by a number with over six hundred digits! How might one choose the one correct order out of so many possibilities?

With paper chromatography, the answer proved easier than might have been expected. Working with the insulin molecule (made up of but some fifty amino acids), the English biochemist, Frederick Sanger (1918–), spent eight years working out the method. He broke down the insulin molecule partway, leaving short chains of amino acids intact. He separated these short chains chromatographically and identified the amino acids making up those chains, as well as the order of arrangement in each. This was not an easy task, since even a four-unit fragment can be arranged in twenty-four different ways, but it was not a completely formidable task either. Slowly, Sanger was able to deduce which longer chains could give rise to just those short chains he had discovered and no others. Little by little, he built up the structure of longer and longer chains until, by 1953, the exact order of the amino acids in the whole insulin molecule had been worked out.

The value of the technique was demonstrated almost at once by the American biochemist, Vincent du Vigneaud (1901–). He applied the Sanger technique to the very simple molecule of "oxytocin," a hormone made up of only eight amino acids. Once their order was worked out, the fact that there were only eight made it practical to synthesize the compound with each of the amino acids in the proper place. This was done in 1954, and the synthetic oxytocin was found to be exactly like the natural hormone in all respects.

Both Sanger's feat of analysis and Du Vigneaud's feat of synthesis have been repeated on a larger scale since. In 1960, the arrangement of the amino acids in an enzyme called "ribonuclease" was worked out. The molecule was composed of 124 amino acids, two and a half times as many as the number of amino acids in the insulin mole-

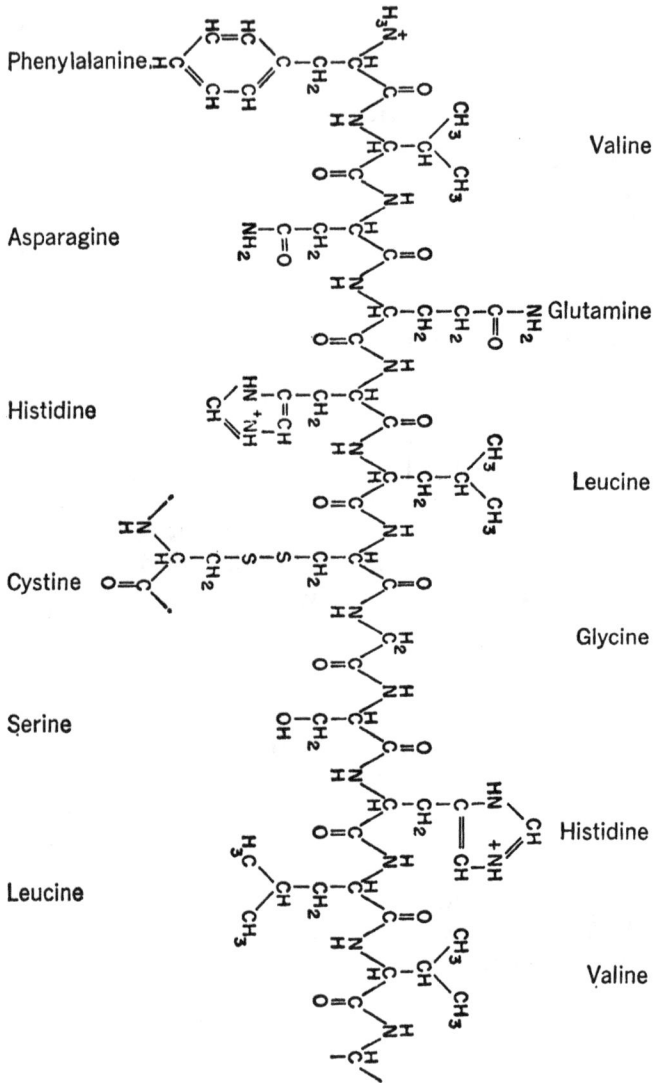

Phenylalanine

Valine

Asparagine

Glutamine

Histidine

Leucine

Cystine

Glycine

Serine

Histidine

Leucine

Valine

cule. Furthermore, fragments of the ribonuclease molecule could be synthesized and studied for enzymatic effectiveness. By 1963, it was discovered in this way that amino acids 12 and 13 ("histidine" and "methionine") were essential for the action of the molecule. This was a long step toward determining the exact manner in which a particular enzyme molecule performed its function.

Thus, as the mid-century progressed, the protein molecule was gradually being tamed by the advance of knowledge.

R - Side Chains.

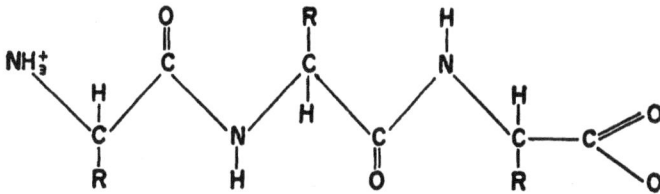

FIGURE 6. Chemical formulas showing the complex structure of a protein. Above is a portion of one of two peptide chains which form the protein molecule of insulin. The peptide backbone is repeated along the center of the chain and a few of the animo acids are shown linked in as side chains. On the facing page is a portion of the peptide chain which forms the backbone of a protein. R represents the amino acid side chains. (After a drawing in *Scientific American*.)

CHAPTER 14

Molecular Biology: Nucleic Acid

Viruses and Genes

But even as the protein molecule came under control, it was suddenly, and quite surprisingly, replaced by another type of substance as the prime "chemical of life." The importance of this new substance made itself felt, first of all, through a line of research brought into play by the question of the nature of the filtrable virus.

The nature of the virus remained a puzzle for a generation. It was known to cause disease and methods were developed to counter it in this respect (see page 135), but the thing itself, rather than merely its effects, remained unknown.

Eventually, filters were developed that were fine enough to hold back the virus and from that it could be estimated that the virus particles, whatever they were, while very much smaller than even the smallest known cells, were still larger than even very large protein molecules. They proved thus to be structures that were intermediate between cells and molecules.

It was the electron microscope (see page 148) that finally revealed them as objects that could be sensed. They proved to cover a large range of sizes, from tiny dots not very much bigger than a large protein molecule, to sizable structures with regular geometrical shapes and with an apparent internal organization. The bacteriophages were among the largest viruses for all that they preyed on such small organisms, and some of them were tailed, like tiny tadpoles. Above the virus range and yet still smaller than even the smallest ordinary bacteria were the "rickettsia" (named for Ricketts [see page 106] because microorganisms of this type caused Rocky Mountain fever, the disease that bacteriologist had investigated.)

The question was thus raised as to whether this group of organisms, which seemed to fill the range between the smallest cells and the largest molecules, were alive or not. A startling development that seemed to militate against the hypothesis that they were alive came in 1935. The American biochemist, Wendell Meredith Stanley (1904–   ), then working with extracts of tobacco mosaic virus, was able to obtain fine needlelike crystals. These, when isolated, proved to possess all the infective properties of virus, and in high concentration. In other words, he had crystalline virus and a living crystal was a concept that was quite difficult to accept.

On the other hand, might it not be conjectured that the cell theory was inadequate and that intact cells were not after all the indivisible units of life. The virus was much smaller than a cell and, unlike cells, did not possess the capacity for independent life under any circumstances. Yet it managed to get inside cells and once there it reproduced itself and behaved in certain key respects as though it were alive.

Might there not be, then, some structure within the cell, some subcellular component that was the true essence of life; one that controlled the rest of the cell as its tool? Might a virus not be that cellular component broken loose, somehow, waiting only to invade a cell and take it over from its rightful "owners"?

If this were so, then such subcellular components ought to be located in normal cells, and the logical candidates for the honor seemed to be the chromosomes (see page 83). In the first years of the twentieth century, it became plain that the chromosomes carried the factors governing the inheritance of physical characteristics and so they controlled the rest of the cell as the key subcellular component would be expected to do. The chromosome, however, was far larger than the virus.

But there were far fewer chromosomes than there were inheritable characteristics, so that it could only be concluded that each chromosome was made up of many units, perhaps thousands, each of which controlled a single characteristic. These individual units were named "genes" in 1909 by the Danish botanist, Wilhelm Ludwig Johannsen (1857–1927), from a Greek word meaning "to give birth to."

In the first decades of the twentieth century, the individual gene, like the individual virus, could not be seen, and yet it could be worked with fruitfully. The key to such work came when the American geneticist, Thomas Hunt Morgan (1866–1945), introduced a new biological tool in 1907, a tiny fruit fly, *Drosophila melanogaster*. This was a small insect, capable of being bred in large numbers and with virtually no trouble. Its cells, moreover, possessed but four pairs of chromosomes.

By following fruit-fly generations, Morgan discovered numerous cases of mutations, thus extending to the animal kingdom what De Vries (see page 79) had discovered among plants. He was further able to show that various characteristics were linked; that is, inherited together. This meant that the genes governing such characteristics were to be found on the same chromosome, and this chromosome was inherited, of course, as a unit.

But linked characteristics were not eternally linked. Every once in a while, one was inherited without the other. This came about because pairs of chromosomes occasionally switched portions ("crossing over"), so that the integrity of an individual chromosome was not absolute.

Such experiments even made it possible to locate the spot on the chromosome at which a particular gene might exist. The greater the length of chromosome separating two genes, the greater the likelihood that crossing over at a random spot would separate the two. By studying the frequency with which two particular linked characteristics were unlinked, the relative positions of the genes could be established. By 1911, the first "chromosome maps" for fruit flies were being drawn up.

One of Morgan's students, the American geneticist, Hermann Joseph Muller (1890–), sought a method for increasing the frequency of mutations. In 1919, he found that raising the temperature accomplished this. Furthermore, this was not the result of a general "stirring up" of the genes. It always turned out that one gene was affected, while its duplicate on the other chromosome of the pair was not. Muller decided that changes on the molecular level were involved.

He therefore tried X rays next. They were more energetic than gentle heat, and an individual X ray striking a chromosome would certainly exert its effect on a point. By 1926, Muller was able to show quite clearly that X rays did indeed greatly increase the mutation rate. The American botanist, Albert Francis Blakeslee (1874–), went on to show, in 1937, that the mutation rate could also be raised by exposure to specific chemicals ("mutagens"). The best example of such a mutagen was "colchicine," an alkaloid obtained from the autumn crocus.

Thus, by the mid-1930s, both viruses and genes were losing their quality of mystery. Both were molecules of approximately the same size and, as it quickly turned out, of approximately the same chemical nature. Could the genes be the cell's tame viruses? Could a virus be a "wild gene"?

The Importance of DNA

Once viruses were crystallized, it became possible to analyze them chemically. They were protein, of course, but a particular variety of protein; a variety called "nucleoprotein." The advance of staining methods made it possible also to work out the chemical nature of individual subcellular structures, and it turned out that the chromosomes, too (and therefore the genes), were nucleoprotein.

A nucleoprotein molecule consists of protein in association with a phosphorus-containing substance known as "nucleic acid." The nucleic acids were first discovered in 1869 by a Swiss biochemist, Friedrich Miescher (1844–1895). They were so named because they were first detected in cell nuclei. Later, when they were found to exist outside the cell nucleus, too, it was too late to change the name.

The nucleic acids were first studied in detail by a German biochemist, Albrecht Kossel (1853–1927), who, in the 1880s and thereafter, broke nucleic acids down into smaller building blocks. These included phosphoric acid and a sugar he could not identify. In addition there were two compounds of a class called "purines" with molecules made up of two rings of atoms, including four nitrogens. These Kossel named "adenine" and "guanine" (and they are sometimes referred to simply as A and G). He found also three "pyrimidines" (compounds with a single ring of atoms, including two nitrogens), which he named "cytosine," "thymine," and "uracil" (C, T, and U).

A Russian-American chemist, Phoebus Aaron Theodor Levene (1869–1940), carried matters further in the 1920s and 1930s. He showed that in the nucleic acid molecule, a phosphoric acid molecule, a sugar molecule, and one of the purines or pyrimidines formed a three-part unit which he called a "nucleotide." The nucleic acid molecule is built up of chains of these nucleotides, as proteins are built up of chains of amino acids. The nucleotide chain is built up by connecting the phosphoric acid of one nucleotide to the sugar group of the neighboring nucleotide. In this way a "sugar-phosphate backbone" is built up, a backbone from which individual groupings of purines and pyrimidines extend.

Levene further showed that the sugar molecules found in nucleic acids were of two types: "ribose" (containing only five carbon atoms instead of the six carbon atoms in the better-known sugars) and "deoxyribose" (just like ribose except that its molecule possessed one fewer oxygen atom). Each nucleic acid molecule contained one type of sugar or the other, but not both. Thus, two types of nucleic acid could be distinguished: "ribosenucleic acid," usually abbreviated RNA; and "deoxyribosenucleic acid," usually abbreviated DNA. Each contained purines and pyrimidines of only four different varieties. DNA lacked uracil and possessed only A, G, C, and T. On the other hand, RNA lacked thymine, and possessed A, G, C, and U.

The Scottish chemist, Alexander Robertus Todd (1907–   ), confirmed Levene's deduction in the 1940s by actually synthesizing various nucleotides.

Biochemists did not at first attach special importance to nucleic acids. Protein molecules were, after all, found in association with a variety of nonprotein adjuncts, including sugars, fats, metal-containing groups, vitamin-containing groups, and so on. In every case, it was the protein that was considered the essential portion of the molecule with the nonprotein section quite subordinate. Nucleoproteins might be found in chromosomes and in viruses, but it was taken for granted that the nucleic acid portion was subsidiary and that the protein was the thing itself.

Kossel, in the 1890s, made some observations, however, which, by hindsight, we can see to be most significant. Sperm cells consist almost entirely of tightly packed chromosomes and carry the chemical substances that include the complete "instructions" by which the father's share of inherited characteristics are passed on to the young. Yet Kossel found sperm cells to contain very simple proteins, much simpler ones than those found in tissues, whereas the nucleic acid content seemed to be the same in nature as those in tissues. This might make it seem more likely that the inheritance instructions were included in the sperm's unchanged nucleic acid molecules rather than in its grossly simplified protein.

Biochemists remained unmoved, nevertheless. Not only was faith in the protein molecule unshakable but, through the 1930s, all evidence seemed to point to the fact that nucleic acids were quite small molecules (made up of only four nucleotides each) and therefore far too simple to carry genetic instructions.

The turning point came in 1944 when a group of men headed by the American bacteriologist, Oswald Theodore Avery (1877–1955), were working with strains of pneumococci (pneumonia-causing bacteria). Some were "smooth" strains (S), possessing an outer capsule about the cell; while others were "rough" strains (R), lacking such a capsule.

Apparently the R strain lacked the ability to synthesize the capsule. An extract from the S strain added to the R strain converted the latter into the S strain. The extract could not itself bring about the formation of the capsule but, apparently, it produced changes in the R strain that made the bacteria themselves capable of the task. The extract carried the genetic information necessary to change the physical characteristics of the bacteria. The totally startling part of the experiment came with the analysis of the extract. It was a solution of nucleic acid and nucleic acid alone. No protein of any kind was present.

In this one case at least, nucleic acid was the genetic substance, and not protein. From that moment on it had to be recognized that it was nucleic acid after all that was the prime and key substance of life. Since 1944 also saw the introduction of the technique of paper chromatography, it might fairly be termed the greatest biological year since 1859 when *The Origin of Species* was published (see page 65).

In the years since 1944, the new view of nucleic acid

has been amply confirmed, most spectacularly perhaps through work on viruses. Viruses were shown to have an outer shell of protein, with a nucleic acid molecule in the inner hollow. The German-American biochemist, Heinz Fraenkel-Conrat (1910–), was able, in 1955, to tease the two parts of the virus apart and put them together again. The protein portion by itself showed no infectivity at all; it was dead. The nucleic acid portion by itself showed a bit of infectivity; it was "alive," though it needed the protein portion to express itself most efficiently.

Work with radioactive isotopes showed clearly that when a bacteriophage, for instance, invaded a bacterial cell, only the nucleic acid portion entered the cell. The protein portion remained outside. Inside the cell, the nucleic acid not only brought about the manufacture of more nucleic acid molecules like itself (and not like those native to the bacterial cell), but also protein molecules to form its own shell, its own characteristic protein, and not that of the bacterial cell. Certainly there could no longer be any doubt that the nucleic acid molecule, and not protein, carried genetic information.

Virus molecules contained either DNA or RNA or both. Within the cell, however, DNA was found in the genes exclusively. Since the genes were the units of heredity, the importance of the nucleic acid resolved itself into the importance of DNA.

Nucleic Acid Structure

After Avery's work, nucleic acids came under prompt and intense study. They were quickly found to be large molecules. The illusion that they were small came about because earlier methods of extraction had been harsh enough to break up the molecules into smaller fragments as they were being extracted. Gentler techniques extracted nucleic acid molecules as large as or larger than the largest protein molecules.

The Austrian-American biochemist, Erwin Chargaff (1905–), broke down nucleic acid molecules and subjected the fragments to separation by paper chromatography. He showed, in the late 1940s, that in the DNA molecule, the number of purine groups was equal to the number of pyrimidine groups. More specifically, the number of adenine groups (a purine) was usually equal to the number of thymine groups (a pyrimidine), while the number of guanine groups (a purine) was equal to the number of cytosine groups (a pyrimidine). This might be expressed as A = T and G = C.

The New Zealand-born British physicist, Maurice Hugh Frederick Wilkins (1916–), applied the technique of X-ray diffraction (see page 158) to DNA in the early 1950s, and his colleagues at Cambridge University, the English biochemist, Francis Harry Compton Crick (1916–) and the American biochemist, James Dewey Watson (1928–), attempted to devise a molecular structure that would account for the data obtained by Wilkins.

Pauling had just evolved his theory of the helical structure of proteins (see page 157), and it seemed to Crick and Watson that a helical DNA molecule would fit in with Wilkins' data. They needed a double helix, however, to account for Chargaff's findings as well. They visualized the DNA molecule as consisting of two sugar-phosphate backbones winding up about a common axis and forming a cylindrical molecule. The purines and pyrimidines extended inward from the backbones, approaching the center of the cylinder. To keep the diameter of the cylinder uniform, a large purine must always be adjacent to a small pyrimidine. Specifically, an A must adjoin a T and a G must adjoin a C and it is thus that Chargaff's findings were explained.

Furthermore, an explanation was now available for the key step in mitosis, the doubling of the chromosomes (and for a related problem as well, the manner in which virus molecules reproduced themselves within a cell).

Each DNA molecule formed a replica of itself ("repli-

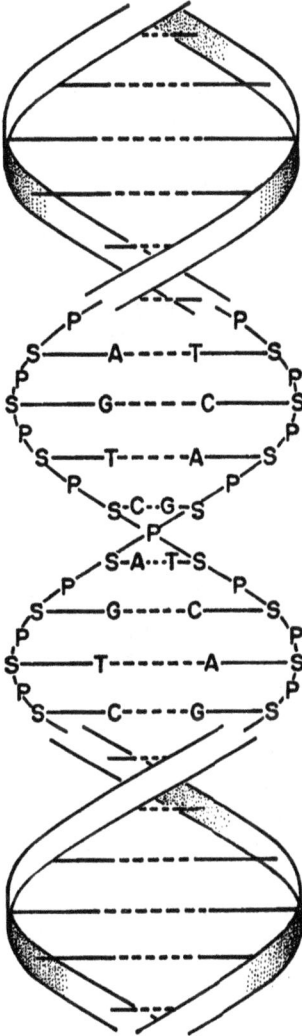

A - adenine   C - cytosine
G - guanine   P - phosphate
T - thymine   S - sugar

FIGURE 7. The double helix of the DNA molecule. The backbone strands are made up of alternating sugar (S) and phosphate (P) groups. Extending inward are the bases, adenine (A), guanine (G), thymine (T), and cytosine (C). The dashed lines are hydrogen bonds which link the strands. In replication, each of these strands will produce its complement from the purines and pyrimidines (A,G,C,T, etc.) that are always present in the cell. (After a drawing in *Scientific American*.)

cation") as follows: The two sugar-phosphate backbones unwound and each served as a model for a new "complement." Wherever an adenine existed on one backbone, a thymine molecule was selected from among the supply always present in the cell, and vice versa; wherever a guanine molecule was present, a cytosine molecule was selected, and vice versa. Thus, backbone 1 built up a new backbone 2, while backbone 2 built up a new backbone 1. Pretty soon, two double helices existed where only one had before.

If DNA molecules did this all along the line of a chromosome (or virus), one ended with two identical chromosomes (or viruses) where only one had existed before. The process was not always carried through perfectly. When an imperfection occurred in the replication process, the new DNA molecule was slightly different from its "ancestor"; and one had a mutation.

This Watson-Crick "model" was announced to the world in 1953.

The Genetic Code

But how did the nucleic acid molecule manage to pass on information concerning physical characteristics? The answer to that was made known through the work of the American geneticists, George Wells Beadle (1903–   ) and Edward Lawrie Tatum (1909–   ). In 1941, they began experiments with a mold called *Neurospora crassa*, one that was capable of living on a nutrient medium containing no amino acids. The mold could manufacture all its own amino acids out of simpler nitrogen-containing compounds.

If the molds were subjected to X rays, however, mutations were formed and some of these mutations lacked the ability to form all their own amino acids. One mutated strain might, for instance, be unable to form the amino acid, lysine, but would have to have it present in the nutrient mixture in order to grow. Beadle and Tatum were able to show that this inability was caused by the lack of a specific enzyme that the ordinary unmutated strain possessed.

They concluded that it was the characteristic function of a particular gene to supervise the formation of a particular enzyme. The nucleic acid molecules passed on in sperm and egg possessed within themselves the capacity of producing a particular set of enzymes. The nature of this set governed the cell chemistry; and the nature of the cell chemistry produced all the characteristics concerning whose heredity scientists inquired. Thus, one passed from DNA to physical characteristics.

The production of enzymes by the genes must, however, clearly be performed through intermediaries, since the DNA of the genes remained within the nucleus while protein synthesis went on outside the nucleus. With the advent of the electron microscope, the cell was studied in new and much subtler detail and the exact site of protein synthesis was found.

Organized granules, much smaller than the mitochondria (see page 148) and therefore called "microsomes" (from Greek words meaning "small bodies"), had been noted in great numbers within the cell. By 1956, one of the most assiduous of the electron microscopists, the Rumanian-American, George Emil Palade (1912–), had succeeded in showing that the microsomes were rich in RNA. They were therefore renamed "ribosomes," and it was these ribosomes that proved to be the site of protein manufacture.

The genetic information from the chromosomes must reach the ribosomes and this was done through a particular variety of RNA called "messenger-RNA." This borrowed the structure of a particular DNA molecule within the chromosomes, and traveled out with that structure to a ribosome on which it layered itself. Small molecules of "transfer-RNA," first studied by the American biochemist, Mahlon Bush Hoagland (1921–), attached themselves to specific amino acids; then, carrying the amino acids, attached themselves to matching spots on the messenger-RNA.

The chief remaining problem was to decide how a particular molecule of transfer-RNA came to attach itself to a particular amino acid. The simplest solution would be to imagine an amino acid attaching itself to a purine or pyrimidine of the nucleic acid; a different amino acid to each purine or pyrimidine. However, there are about twenty different amino acids and only four purines and pyrimidines to a nucleic acid molecule. For that reason, it seems clear that a combination of at least three nucleotides must be matched to each amino acid. (There are 64 different possible combinations of three nucleotides.)

Matching the trinucleotide combination to the amino acid has been the great biological problem of the early 1960s and this is usually referred to as "breaking the genetic code." Men such as the Spanish-American biochemist, Severo Ochoa (1905–), have been active in this respect.

The Origin of Life

The advances made in molecular biology in the mid-twentieth century have brought the mechanist position to an unprecedented pitch of strength. All of genetics can be interpreted chemically, according to the laws that hold for animate and inanimate alike. Even the world of the mind shows signs of giving way before the torrent. It would seem that the process of learning and remembering is not the establishment and retention of nerve pathways (see page 122), but the synthesis and maintenance of specific RNA molecules. (Indeed, flatworms, a very simple form of life, have been shown capable of learning tasks by eating other flatworms that had already learned the tasks. Presumably, the eater incorporated intact RNA molecules of the eaten into its own body.)

That left the one facet of biology that had represented a clear victory for the nineteenth-century vitalist position

—the matter of the disproof of spontaneous generation (see page 92). With the twentieth century, that disproof had grown less attractive in the absolute sense. If, indeed, life form could *never* develop from inanimate manner, then how did life begin? The most natural assumption was to suppose that life was created by some supernatural agency, but if one refused to accept that, what then?

In 1908, the Swedish chemist, Svante August Arrhenius (1859–1927), speculated on the origin of life without invoking the supernatural. He suggested that life had begun on earth when spores reached our planet from outer space. The vision arose of particles of life drifting across the vast reaches of emptiness, driven by light pressure from the stars, landing here and there, fertilizing this planet or that. Arrhenius' notion, however, merely pushed back the problem; it didn't solve it. If life did not originate on our own planet, how did it originate wherever it did originate?

It was necessary to consider once again whether life might not possibly originate from nonliving matter. Pasteur had kept his flask sterile for a limited time, but suppose it had remained standing for a billion years? Or suppose not just a flask of solution had remained standing for a billion years, but a whole ocean of solution? And suppose that the ocean might be doing so under conditions far different from those which prevail today?

There is no reason to think that the basic chemicals of life have changed essentially, over the eons. It is quite likely, in fact, that they have not. Thus, small quantities of amino acids persist in some fossils that are tens of millions of years old and those that are isolated are identical to amino acids that occur in living tissue today. Nevertheless, the chemistry of the world generally may have changed.

Growing knowledge of the chemistry of the universe has led men such as the American chemist, Harold Clayton Urey (1893–), to postulate a primordial earth, in which the atmosphere was a "reducing" one, rich in hydrogen and in hydrogen-containing gases such as methane and ammonia, and with free oxygen absent.

Under such conditions there would be no ozone layer in the upper atmosphere (ozone being a form of oxygen). Such an ozone layer now exists and absorbs most of the sun's ultraviolet radiation. In a reducing atmosphere, this energetic radiation would penetrate to sea level and bring about reactions in the ocean which, at present, do not take place. Complex molecules would slowly form and, with no life already present in the oceans, these molecules would not be consumed but would accumulate. Eventually, nucleic acids complex enough to serve as replicating molecules would be formed and this would be the essential of life.

Through mutation and the effects of natural selection, more and more efficient forms of nucleic acid would be produced. These would eventually develop into cells, of which some would begin to produce chlorophyll. Photosynthesis (with the aid of other processes not involving life, perhaps) would convert the primordial atmosphere into the one with which we are familiar, one rich in free oxygen. In an oxygen atmosphere and in a world already teeming with life, spontaneous generation of the type just described would then no longer be possible.

To a very great extent this is speculation (although carefully reasoned speculation), but, in 1953, one of Urey's pupils, Stanley Lloyd Miller (1930–   ), performed what has become a famous experiment. He began with carefully purified and sterilized water and added an "atmosphere" of hydrogen, ammonia, and methane. He circulated this through a sealed apparatus past an electric discharge which represented an energy input designed to mimic the effect of solar ultraviolet. He kept this up for a week, then separated the components of his water solution by paper chromatography. He found simple organic compounds among those components and even a few of the smaller amino acids.

In 1962, a similar experiment was repeated at the University of California, where ethane (a two-carbon compound very similar to the one-carbon methane) was added to the atmosphere. A larger variety of organic compounds was obtained. And in 1963, adenosine triphosphate, one of the key high-energy phosphates (see page 146) was synthesized in similar fashion.

If this can be done in a small apparatus in a matter of a week, what might not have been done in a billion years with a whole ocean and atmosphere to draw upon?

We may yet find out. The course of evolution, pushed back to the dawn of earth's history may seem difficult to work out, but if we reach the moon we may be able more clearly to make out the course of chemical changes prior to the advent of life itself. If we reach Mars, we may even (just possibly) be able to study simple life forms that have developed under conditions quite different from those on earth, and this, too, may be applicable to some of our earthly problems.

Even on our own planet, we are learning more each year about life forms under the alien conditions of the oceanic abysses, for in 1960, men penetrated to the very bottom of the deepest of these. It is even possible that in the ocean we may establish communications with non-human intelligence in the form of dolphins.

The human mind itself may yield its secrets to the probings of the molecular biologists. Through increasing knowledge of cybernetics and electronics we may be able to develop forms of inanimate intelligence.

But why guess when we need only wait? It is perhaps the most satisfying aspect of scientific work that no matter how great the advances or how startling and smashing the gains of knowledge over the unknown, what remains for the future is always still greater, still more exciting, still more wonderful.

What may not yet be revealed during the very lifetime of those now living?

Index

www.ingramcontent.com/pod-product-compliance
Lightning Source LLC
Chambersburg PA
CBHW050459190326
41458CB00005B/1348